SEASON OF CANCER

BY

Pero Metkovic

Are the diseases associated with rotation of Earth? Are they related to the influence of celestial bodies? Was astrology successful in medicine in old age? Is the science of Old Babylon acceptable to us?

Through history, flu has usually started in China. What for...? Earth rotates from east to west. In front of China was the Pacific Ocean. That was the reason why every flu started in China.

Rotation of earth is directly related to the disease. Every flu has started in China. Now we already have two parameters, to study the cause of the flu, the rotation of the earth and the position of the celestial bodies compared to earth.

Astrologers used to measure the position of celestial bodies to determine how much the disease would be strong in a given period of time. Illness caused by the influence of celestial bodies is mostly manifested as a flu. The most important space effect on our bodies is the sun.

What is flu? Inflammation usually raises our body temperature. Body temperature is one of the parameters of influenza. We feel tired, nervous. We vomit. Sometimes we want to sleep. Our body is not the same as before. It has changed.

If the astrologers were still working, they would say that an upcoming flu would be so and so strong. It should have such and such characteristics.

Taking all these elements into consideration, we will find that something affects our body. For example, cold weather will affect our bodies. We shall get sick and get cold. In this case, if we breathe the cold air, we shall get sick. If we were not warmly dressed, we should be sick.

Here we see the direct impact of weather on our body. Our new illness is a result of cold weather, in fact the contact of the cold on the organs of our body. Suppose we fall into cold water in the winter. Again we will become sick or cold. Contact of our body with cold water will cause our new illness.

Here we can check our contact with the heat. If our body touched some hot object, for example a hot stove, we would hurt that part of our body. We would get burns on that part of the body that was touching the hot stove.

If someone falls in a very hot water he will get burns on his body. Our body temperature will increase. We will feel nervous and tired. Our body will become different than it was. In all these cases, exactly the same thing will happen. We will get sick. Our body will change.

Our contact with the cold will be visible on our skin. Our contact with the cold could be seen on our skin, but we do not usually see it. Our heat contact, like burns, is easy to see.

Our temperature will rise. We will have more symptoms, such as tiredness, etc. Doctors will find viruses in our blood.

What actually happened? How did viruses enter our blood? Our body has created viruses. In fact, viruses are created in our body. Space influence or cold has damaged our organs, as the sensitive mucous membranes of the mouth, nose and respiratory organs.

Injury on our body will cause swelling and inflammation on the skin and nearer layers of the body. Viruses are the result of injuries and inflammations that occurred at that place of the body. This injury caused the inflammation. Newborn viruses are the result of inflammation. In such damaged skin and tissue, which is overwhelmed and partially damaged, viruses are produced. They are creatures that arise under such body conditions.

Each effect produces different mutations of the viruses. Viruses are secondary microorganisms in fact a consequence of cold.

This applies to the weather's cold but there are various influences on the human body. We can call them heavenly influence.

The influence of weather can be defined as the cold that injures our body. In contact with such weather organs of the body as a mucous membrane or lung can be injured.

Celestial influences occur in certain periods of the year. They are linked to the Earth position in such periods of the year. In Dubrovnik, Croatia, February is the month when people get inflammation of the throat.

So many factors can affect our body. Some of them we can identify and some not. Because of this, our body is sensitively. Some of these factors will affect our body's illness. Cold will attack the mucous membrane of our airways. It will become red and swollen. Such a diseased mucous membrane will produce viruses on its surface.

In southern Croatia on the coast, people are sickening if the south wind blows. Although the southern wind is hot it will make our body sick. We will catch the cold and cough. We can hardly recognize all the factors that can affect our body.

We can explain the influence of cold weather. We can hardly explain the illnesses we get due to the deployment of the celestial body. The disease that is formed under the influence of the southern wind "Siroko" is more psychological. Such an effect can produce cough and illness.

The human body responds to impacts. It needs to get used to the cold. We have to wear warm clothes for this, otherwise we could not survive, on the contrary animals can withstand any cold if they live in such a territory.

Earth rotates from west to east. Influence always begins in China, which is first land after the Pacific Ocean. When Earth is in a certain position in the celestial space, then the influence of other bodies may be stronger or weaker. In fact, according to my theory, Earth makes precession over time. Its position is not fixed in the celestial space. Earth does not rotate around the sun according to my book "Formation Of Planet".

Here we see that the influences of other bodies will touch Earth first in the east as in China. Japan is north of China and the impact effect will be smaller. Equatorial territories are most exposed to influences, because Earth is the largest or most fat there.

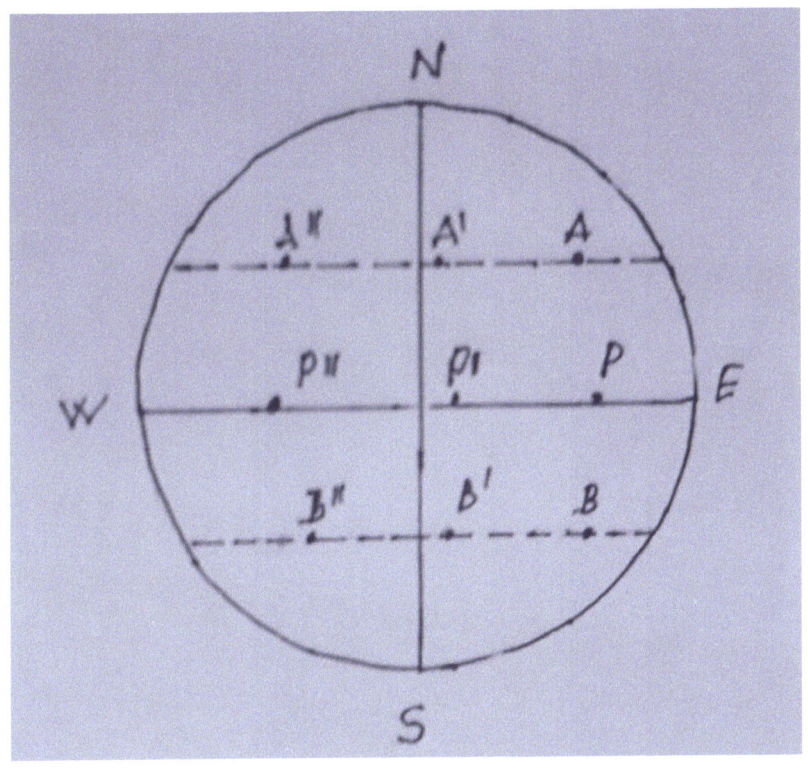

Different motions of points on Earth.

P, P',P" - points on the eqvatory circle.
A, A', A" - points at the 45-degree parallel, north.

B, B', B" - points at the 45-degree parallel, south.

Equatorial circle E - W will make the longest way during Earth rotation. Points P, P 'and P "will receive the most influence from other body emissions because they are most exposed.

The 45-degree parallel to the north and south, in fact, points A, A ', A' on the northern hemisphere and B, B ', B' on the southern hemisphere receives fewer emissions from the celestial bodies. The sun has the highest emissions on Earth, while the moon and the other bodies are weaker.

Point of North Pole "N" and South Pole "S" receive zero or minimum emission from celestial bodies.

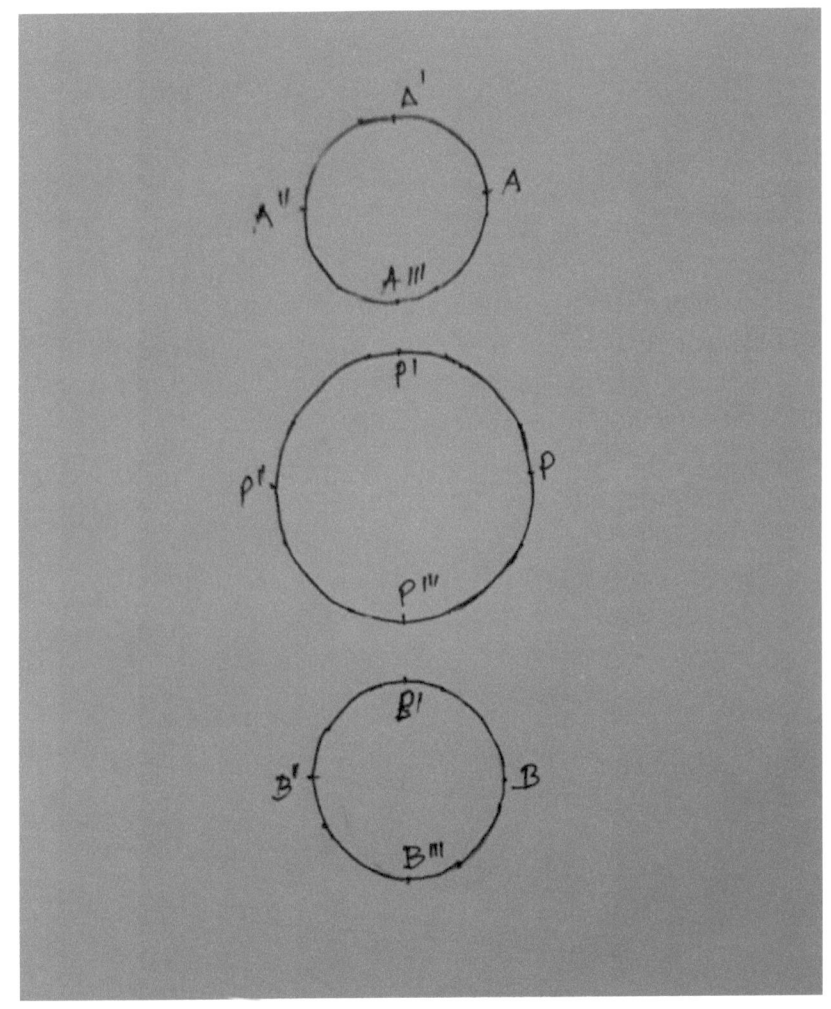

Earth circles of Cellestial bodies emission receipt

P, P', P", P''' are points on Equator.
A, A',A",A''' are points at 45 deg. Circle on North hemisphere.
B, B',B",B''' are points at 45 circle on South hemisphere.

N – Point of North Pole.
S – Point of South Pole.

In this sketch we can observe the points P, P ', P ",P'" make the most way, during one Earth's rotation. The points on the eqvator are the most exposed points on Earth, to the energy emission of celestial bodies.

Points A , A', A ", A '", at 45 deg. latitude circle on the Northern Hemisphere and point B, B ', B ", B'" at 45 deg. latitude circle on the Southern Hemisphere are less exposed to the emission of celestial bodies. These two circles pass a smaller way than the equator during an Earth rotation.

The points on the Poles north and south do not pass any way during an Earth rotation. The poles are least exposed to the emission of celestial bodies.

The surface between the equator and the 45-degree latitude circle of the north and south is most exposed to the emission influences from the celestial body.

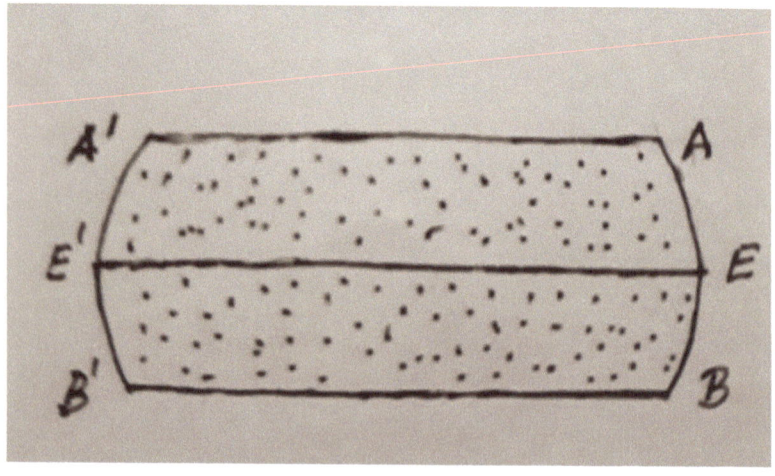

The most exposed area of the celestial bodies emission.

E, E' – Equator
A, A' - A 45-degree latitude circle in the north.
B, B' - Circle of latitude 45 degrees south.

In this sketch we can see the Earth area that is most exposed to the emission of celestial bodies. This is the area between the equator E and E 'and the latitude 45 degrees north and south, A, A' and B, B '. This area receives the highest emission of celestial bodies. This is the area with the greatest potential for influenza illness associated with the emission of celestial bodies. This area we can call the "Equatorial area of the celestial bodies".

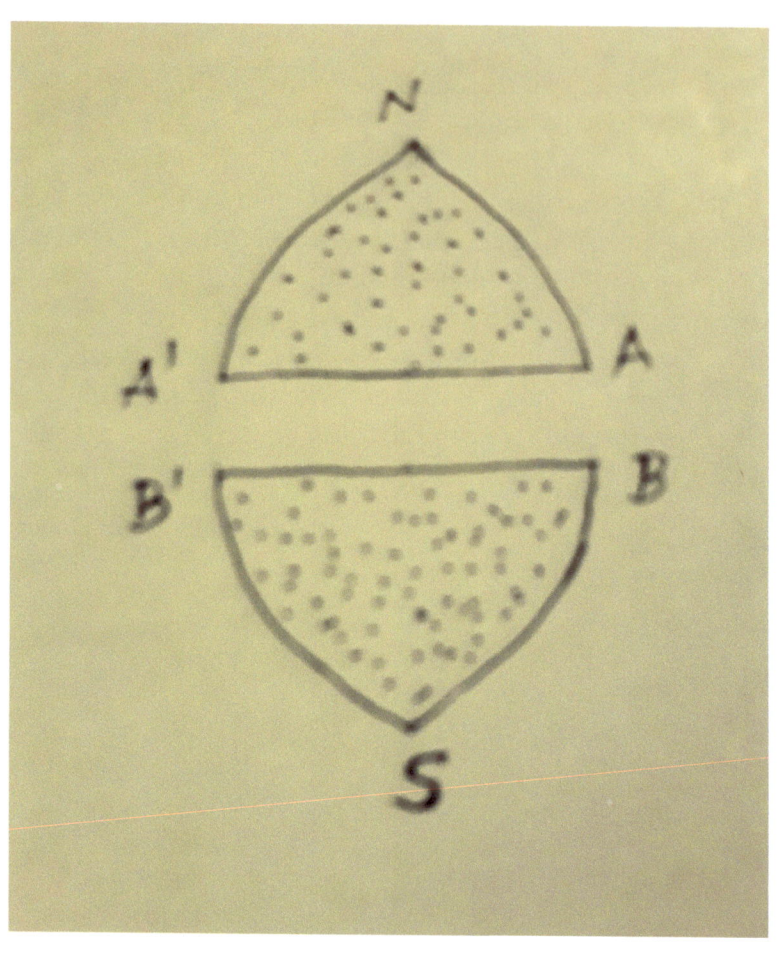

Northern and southern areas of the influence of the emission of celestial bodies.

A, A '- A 45 - degree circle of northern latitude.
N - North point.
B, B '- A 45 - degree circle of southern latitude.
S - South Point.

In this sketch we can see the north and south areas of the 45-degree latitude to the N and S points. These areas are less influenced by celestial bodies emission.

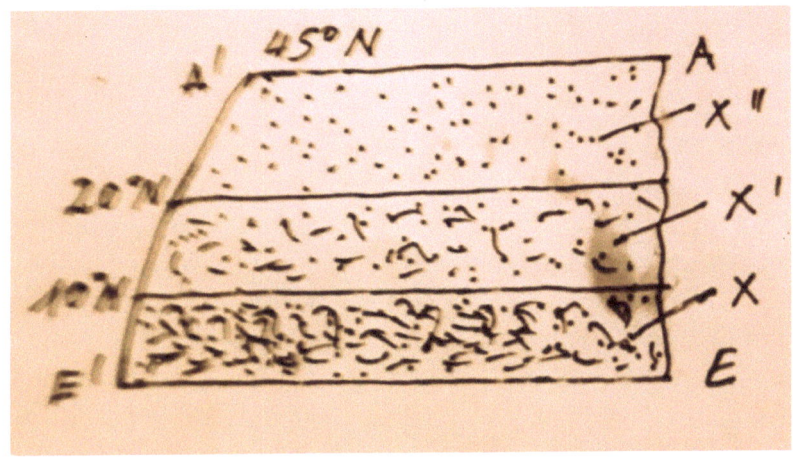

Areas of the greatest influence from the emission of celestial bodies.

X - The area of the equator to 10 degrees north.
X '- Area of 10 to 20 degrees north.
X "- Area of 20 to 45 degrees north.

In this sketch, we can see a segment of "Equator Area Receiving" from 0 to 45 degrees north latitude. The highest reception occurs in the X range of 0 to 10 degrees of latitude. Area X 'receives less emission from the celestial bodies. Area X "receives the smallest emission of celestial bodies.
It is the same on the southern hemisphere. Area X is the largest emission reception zone, which receives more emission than X 'and X".

I have shown here an area on earth that is most exposed to the celestial emission. The sun is the strongest source of celestial energy emissions. We can mark this with number one (1). That is what we can define. We see the sun and we can estimate its brightness and warmth. In fact, the sun emits another kind of energy that I explained in the book "Formation of Planet". Its energy is cold before touching Earth's atmospheric shell. Particles of the atmosphere take on energy, and transmit it to Earth's surface.

The second source of energy is the moon. According to my explanation in the same book, the moon emits energy. We can mark the moon with number two (2). Planets are the third source of energy emissions. According to my explanation in the mentioned book, the planets emit energy. The fourth source of energy emission are stars. Here we can show the energy emission table.

1. Sun
2. Moon
3. Planets
4. Stars
5. Celestial unknown energy

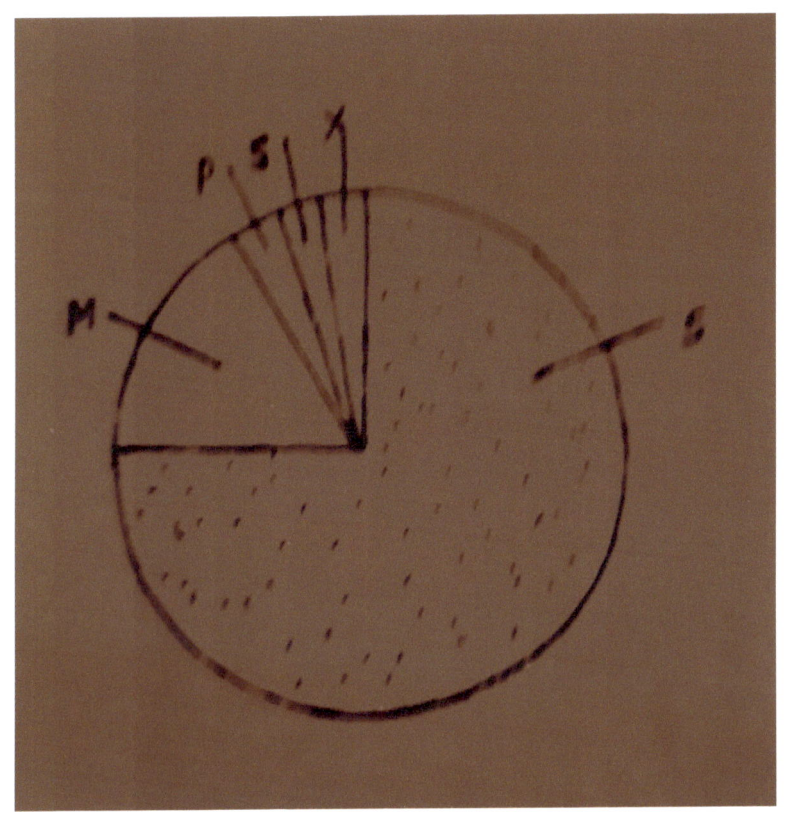

Percentual emission of energy.

S – Sun
M – Moon
P – Planets
S – Stars
X - Celestial unknown energy

According to the sketch, the sun (S) emits at least 75 percent of all celestial emitted energy reaching Earth. The moon (M) emits less than 5 percent. Planets (P) emit less than 1 percent. The stars (S) also emit less than 1 percent. Unknown sources of energy emit less than 2 percent.

Sun (S) is the strongest source of energy that affects the Earth's surface. The sun warms and illuminates the Earth. Life on Earth depends on the sun. Sun rays are medicine to survive any form of life. Although today's medicine negates the medication of the sun's rays, I believe they are healthy. The more we expose them, we will be healthier.

Ancient astrologers linked the Earth's disease to the Earth's position in relation to other celestial bodies. This means that the Earth did not receive the same amount of emissions over tens or hundreds of years. Astrologers counted what the disease would be, whether it would be weaker or stronger. In fact, astrologers counted on the power to receive energy from the celestial bodies.

I have already said that any reception of celestial energy is positive. The Earth is surrounded by a cold. Celestial energy, especially the sun's emission, illuminates the earth. Sun rays penetrate through the cold that wraps the Earth and form an oasis of life in the cold of the earth's surface.

Astrologers were able to measure the power of receiving energy to the Earth's position in relation to the sun. They could say strength and other characteristics of the disease during the year with regard to the observation of celestial bodies. In the last sketch we can see the symbol X, which shows the reception of energy from undefined sources, such as invisible planets.

****

What is the disease actually? If a man comes into contact with the strong cold he can die. Before death he will be sick. How long will his illness last? Will it be one or ten days? In fact we got sick because of the contact with extreme cold. Our body got a strong cold or inflammation.

What happened in the human body because of the illness he got. We can compare man and wolf. They found themselves in extreme cold. During the storm the wind blew so hard. It was snowing. The wolf survived even though it was snowed. It was exposed to storm for a few days.

One man was just a day in the storm. He had a good coat, cap and boots. The next day the safety patrol found him. He was ten days in hospital before he died of cold. Now we can analyze what actually happened to that man. Only a few injuries were found on his skin due to the cold. Which part of the body was attacked by coldness?

We can analyze what actually happened to a man. His skin was mostly healthy. Doctors found only a few injuries on his skin caused by cold. Which part of the body was attacked by coldness?

If we look at the whole body then in this case we can say that the skin was only 5-10 percent attacked and damaged. All illness and injuries occurred in the interior of the body under the skin. The human body reacted against the cold with its entire mass. Nerves, muscles and other parts of the body were involved in the fight against the cold. Eventually, a man died of a heart attack.

In this case, man could die during the first night, in the snow. He somehow managed to survive that night. He could die the next few days, but he managed to survive for ten days. The strength and ability of his body have overpowered the cold.

Cold is one of the influences that can attack our body. Strong coldness can cause various injuries and diseases of our body. We can get pneumonia, high temperature, cough, inflammation of the throat and nose.

What would happen, if astrologers said that this year flu, and the inflammation of the throat and nose would be weaker? What I believe is that Earth will be differently illuminated by sun than last year. This is one explanation, but there may be more.

According to my previous writing Earth or planets do not circulate around Sun, nor does the Moon circulate around Earth. Earth makes some precessions, which change its angles of energy reception and distance from Sun. Because of this Sun is sometimes closer or more distant from Earth. I believe that astrologers measured the angles of receiving solar energy to Earth and its power.

Solar energy gives us life. Strong sun rays eliminate the cold and make us healthier. This will reduce illnesses that occur during winter, as colds and the like. The moon and other planets do not have such an effect on Earth as the Sun.

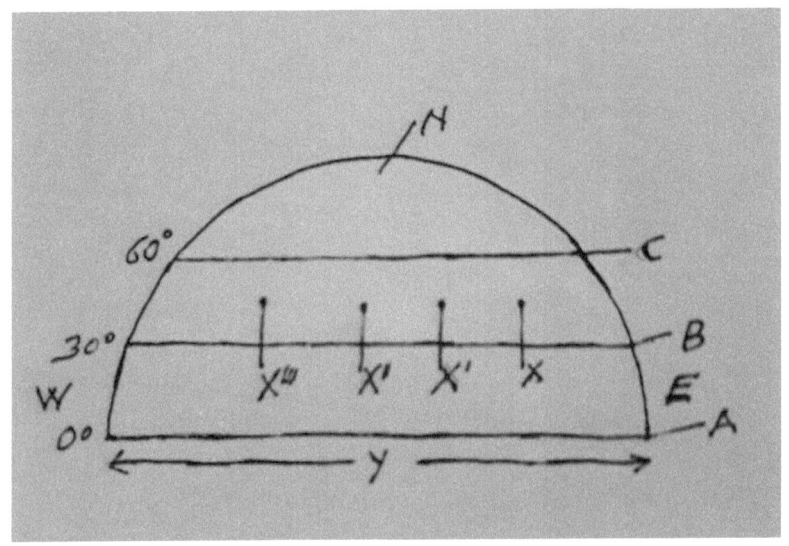

Expansion of influence

Y – Earth's length on the equator.
E – East
W – West
A - Point on the equator
B – Point at 30 degrees north.
C – Point at 60 degrees north
N – North pole
X, X', X"... Points of Influence expansion.

On the sketch we can see the zone between the 30-degree parallel and the 60-degree parallel north. The points X, X ', X"... are instances of influence. As mentioned above, the influenza will begin in China at point X. The first case of influence was observed at point X.

The following case of influence X 'was observed after three months again in China. The third case of influence was observed again after three months in position X″ in India. The next case of influence was observed in position X‴ in Europe, again after three months.

It took around the year that influenza, which started in China in position X, came to Europe in position X‴. What actually happened? Today's medicine considers that the flu, first observed in China, came to Europe for a year by conveying human-to-human infection. In fact, this did not happen in this way. A special celestial factor has caused influenza, which we can call "Factor X".

"Factor X" can be an anomaly that affects the Earth's living world, which may be different from the angle of the solar energy. It can be some radiation from other celestial bodies." Factor X" can be insufficient solar energy that illuminates living creatures.

What actualy happened in position "X" in China. Human beings who were accustomed to a variant of celestial energy remained without it. Influence took the body. In fact, the body became weak without the necessary "Factor X", which was an integral part of the sun's energy reception. All people in the vicinity will get influenza if the required "Factor X" disappears as part of the sun's energy reception in that area.

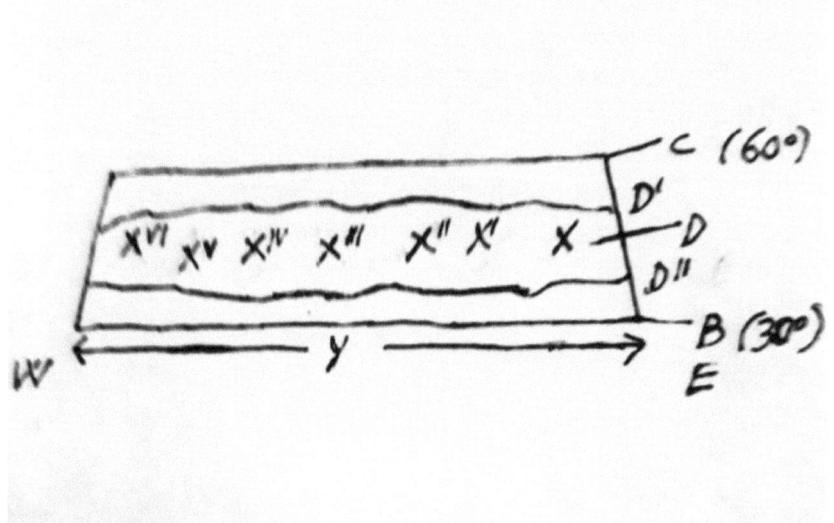

The flu area

Y - Area from East China to Western Europe at 30 degrees north latitude.
B - Point at 30 degrees north latitude.
C - Point at 60 degrees north latitude.
D - Flu area
D', D'' - The boundaries of the flu area.

In this sketch we see the area of spreading the flu (D). In this area, "Factor X" will influence the emergence and spread of the flu. It is bordered by D 'and D ". In this area, "Factor X" will cause flu spread.

How long will the flu spread? Will it be a year, two or more? The duration will be as long as "Factor X" is active. We can try to find out what is "Factor X". Is it a lack of necessary solar energy, or an inadequate angle of invasion of the sun's rays to

Earth. This is something we need because our body depends on sun and other celestial energy. In this case the patient misses something he usually gets from the celestial emission of energy.

In fact, this process started long before we noticed it. We can say that celestial emission began a year before the first symptoms observed in China. The flu came to Europe after another year. This went off in Europe after another year. In China, the flu disappeared after a year, while in Europe it was still effective.

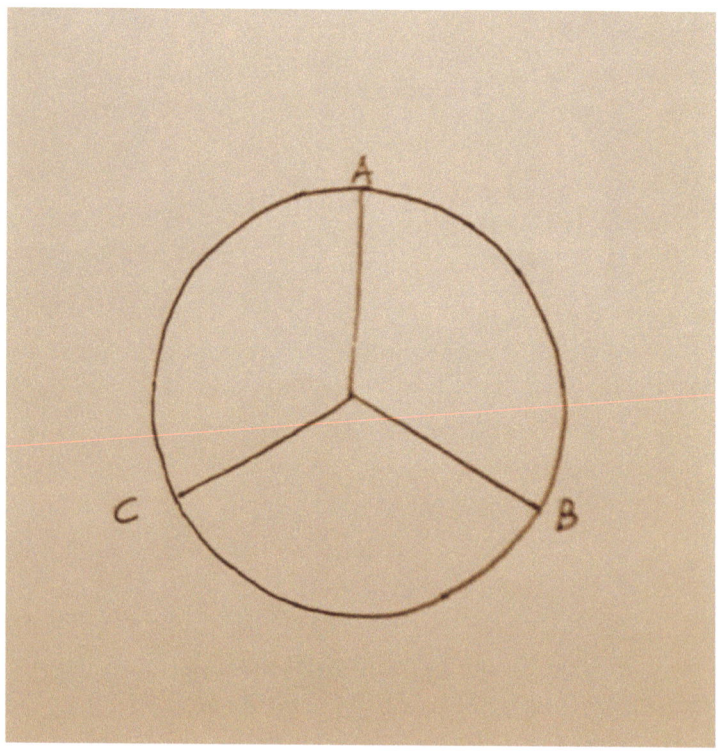

Flu spread map.

A - The time when the celestial emission changed and the time when the emission stopped.

B - The time when the first flu symptoms were seen in China.

C - The time when the flu was observed in Europe after one year.

At sketch, point (A) shows when celestial emission has changed. The sun's rays, which lost "Factor X", began to illuminate Earth. In fact, the celestial emission that has lost its "Factor X" begins to illuminate Earth.

After a year of the emission, the first signs of flu have appeared in China. It is logical that the first symptoms of flu appear in China due to the rotation of the Earth towards it. It took a year for the first symptoms of flu to appear in Europe in Paris, France.

Medicine has explained this case as an infection from man to man. In this case, flu has spread from China to Europe for one year. It is logical that diseases such as this flu first emerge in China. Earth turns east toward China.

In this explanation, people who lost "Factor X" from the emission became weak and sick. Which emission elements are lost I can not say, but these elements are needed for life.

In fact, we live from the sunlight. I emphasize the sun here because it is the main source of the emission. Moon, planets and stars are weaker emission sources. The sun's emission is our other food, but this emission has its diversity. It is sometimes stronger and sometimes weaker. The angle of the sun's rays may be changeable. In addition to the sun we have moon, planets and stars that emit energy to us.

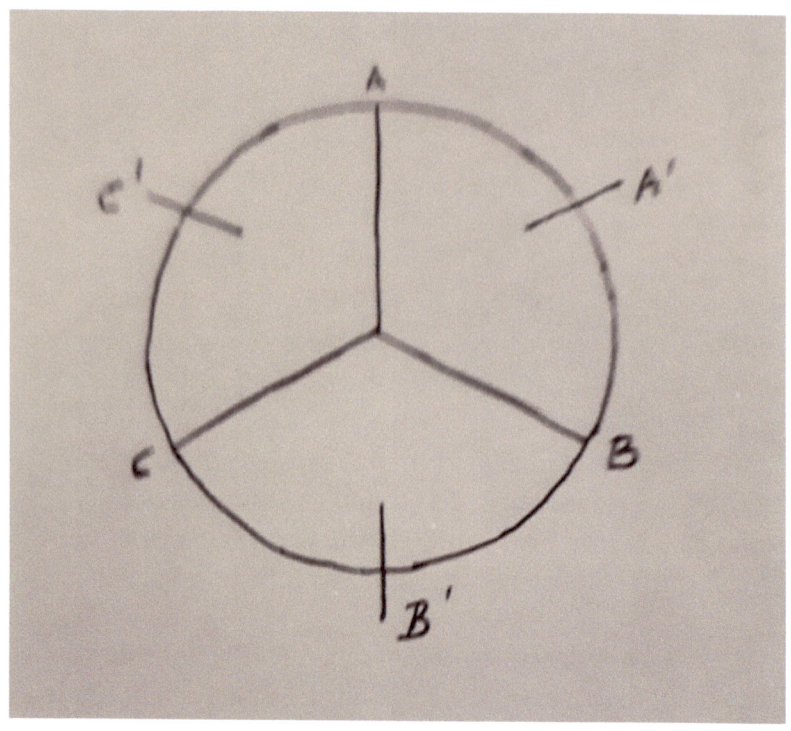

Chart of Influence duration

A - The time when the "Factor X" began to miss in the Sun's emission..
A'- Emission Zone without "Factor X"
B - The time when the first signs of flu were observed in China.
B'- The period of flu, which reached from China to Paris in France.
C - The time when the flu was noticed in Paris in France after one year.
C'- The period of the flu until its disappearance in point A after one year.

In this sketch we can see an approximate explanation of what happened when the flu spread from China to Paris in France.

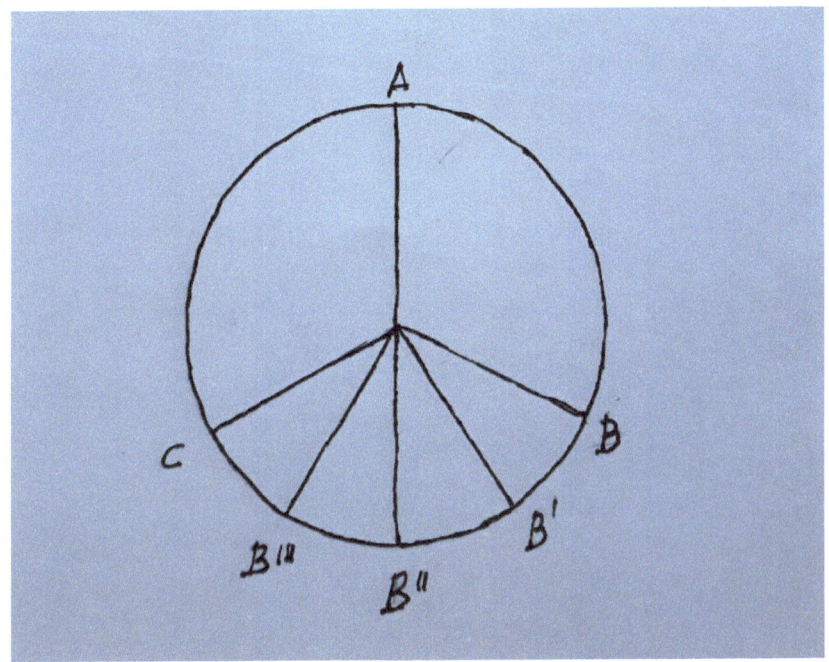

The spread of the flu

A - The time where the solar energy emission lost "Factor X".
B - The time when the first signs of flu were noticed.
B'- The time when the flu was noticed in West China.
B''- The time when the flu was observed in India.
B'''- The time when the flu was observed in Eastern Europe.
C - The time when the flu was noticed in Paris, France.

On the sketch we can see that in the zone between points A and B the celestial emission was carried out without "Factor X" in the period of one year. The first cases of flu were shown in point B after one year. Point C is the time when the flu has reached the latest.

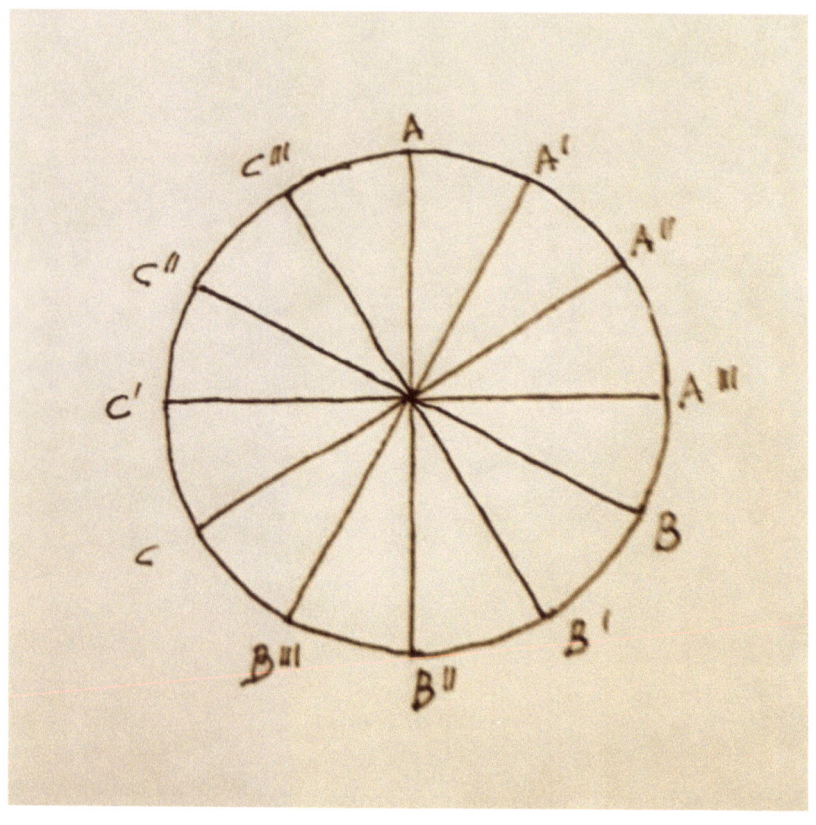

Flu spread division

A - The time where the solar energy emission lost . "Factor X".
A', A'', A''' - Points of a three-month period of emission.
B - The time when the first signs of flu were observed in China.
B', B'', B''' - Points of flu spreading through China, Asia Minor, India to Europe.

C - The time when the flu reached the last point in Paris in France.
C', C'', C''' - The time when the flu began to disappear, first in China C', Asia Minor and India C'' and in Europe C '''.

"Factor X " is a change in celestial emission, in fact in the emission of energy from the sun, moon, planets and stars. Celestial emissions can lose some of their elements or gain new ones. " Factor X" is actually a change of emission, which can influence, or kill man, animals and other living species on Earth.

"Factor X" can kill a living nature like an explosive. This is just an explanation of how "Factor X" works against living nature. This is just a description to explain how "FX" works towards living nature. This is just a description to explain how "Factor X" works towards living nature.

What is actually "Factor X"?

Celestial emission is a factor of life on earth. Life on Earth would be dead without it, but at the same time it could kill us. Are our bodies capable of receiving it?

Many diseases have attacked man through history, which have disappeared, and new ones have emerged. I remember skin ulcers. What happened to them? Have they changed in another form?

In fact, the illnesses arise and after a period of time they disappear. What is the reason for this case?

In this case, skin ulcers disappeared, most likely they changed to some other form. It seems that some diseases are the cause of time periods. Earth changes its position in relation to other celestial bodies. Such Earth precessions causes diseases and their modifications.

What is a disease actually. If we look at the cold, we will see that the human body is attacked by the coldness of weather.

This has caused our body to change. Actually, our body wanted to accept the coldness of weather, but it caused the disease.

For example, the case of skin ulcers, which existed due to "Factor X". In this case, "Factor X" has created skin sores just as cold weather created a cold or flu. We can mark the cold weather as a cause for cold or flu. We cannot find a reason for skin sores. We can say that "Factor X" was the reason for their existence.

Today, cancer is a common disease. Can we say that "Factor X" is the reason for that?

In the example of cold we know the cause of the disease. We can easily take steps to protect ourselves. We can wear warm clothes, etc. How can we protect ourselves from cancer?

We can talk about the period of illness, which we can call "Period X". This period is the duration of a particular disease, or how long this disease will last on Earth. Here we can mention the period of skin ulcers, which lasted so many years, but they disappeared.

The period of illness depends on Earth Precession and "Factor X". In fact, our body is weak and we feel that sickness. Cancer is a common disease today. What is the reason for that? There is a disease caused by "Factor X" in "Period X". Cancer disease will not exist forever. The period of time will only be" Period X" of this disease.

Therefore, through time periods, the disease is changed. They disappear or change into another disease, i.e. they take on a new form, which depends on the relation between Earth and the celestial bodies. In fact diseases are the reaction of our body to external factors.

We can easily recognize the cold and say that the attack of cold will kill us. Our body will react to the cold and may save us. What is it, what does the body do against the cold? Our nerves and other sensitive organs like the skin feel cold. In the event of a cold, the nerves will send a message to all the internal organs, but our entire body with all its organs will not be able to fight the cold and defeat it. The heart will stop and man will die.

Skin ulcers are caused by "Factor X", which affects the mechanism of our body, such as digestion, bloodstream, etc. In other words, ulcers are caused by such sun radiation that creates the possibility of their emergence. This time period of "Factor X" impacts can be termed " Skin Ulcer Season".

Our body and organs work under certain influences. Some of these impacts can be determined and some not. Celestial influence depends on the position of celestial bodies. The influence of the cold weather can be defined with our knowledge and acquired experience.

In southern Croatia, when the southern wind blows, we can easily get cold. Although the wind is not cold, it may even be hot, but we can get sick. We can chill nose, throat, cough etc. In these cases we can identify the causes of inflammations and diseases. People feel depression and headaches during the southern wind, called "Siroko". Such weather affects our nerves.

During the illness, the human body produces viruses and bacterias, that live on our mucous membrane, lungs, throat, etc. These microorganisms are products of disease, and they do not come from the air, as the science teaches. They are a disease product, and if they go in the air, they will die. Even if we move them to another man, with a hand or a cough, they will die. Bacterias and viruses are the product of a certain, sick man.

If we have a cold or flu, a lot of viruses appear on our mouth, throat, nose or lungs. These viruses are the product of our body's disease, but they can not harm another man. Therefore, all microorganisms are not equal. Some of them can be transferred and lived on another body, and others can not.

According to this, every organism can not be transferred and lived on the other body. Microorganisms that can not be transmitted from the body to the body are a consequence, not the beginning of the disease.

Microorganisms of venereal diseases can be transmitted to other bodies and cause the disease there. The disease of the cold produces many microorganisms, but they can not live on another body. Accordingly, these diseases are not contagious. They are the result of cold, or other unknown inflammations due to weather or celestial reasons, as the radiation of celestial bodies.

Coldness, flu, lung inflammation, or throat inflammation are diseases that are not transmitted from man to man. We can compare them with burns to make this explanation look scary. Many microorganisms can be found on the skin where burns occur. These microorganisms are not dangerous to other people.

In fact wounds are a great indicator of microorganisms. If the chemical damages our skin, microorganisms will appear in this area.

Our body can be a habitat for some microorganisms where they will live. This is a case of venereal diseases that live within our sex organs.

One cat was killed in the bush near my home. I was surprised when I saw it. Her entire body was covered with worms. This was not the only time I saw the worms on the cat's creped body. Where did the worms come from? I believe they were born on a cat's body.

In the case of a wound we can identify microorganisms in the immediate area, which originated from our body. They did not come from the air or from anywhere else. Did they live in our body? The wound produces microorganisms from the tissues of our body. This is definitely because where else would they come from. In short, wound is such a body mechanism that can produce microorganisms.

If we get cold, our mucus will become red and bloated, which will become some kind of wound. Microorganisms will be shown in the area of inflammation, which will exist throughout the inflammation period. They will multiply and create microorganisms of the same species. Finally, when inflammation or wound healing microorganisms will creep and disappear. Here we have seen that microorganisms depend on inflammation, in other words about the type of inflammation. The inflammation may be different, and will form new types of microorganisms, depending on the origin of the inflammation. Inflammation of our mucous membranes may occur due to cold, or other meteorological causes, or may be of celestial origin, i.e. due to the diversity of solar energy or the energy of other celestial bodies.

We can classify infectious microorganisms in group (A). They can be transmitted from the body to the body. They usually live in a certain area of the body. Group microorganisms (B) can not usually be transferred from one to another body. These microorganisms will disappear when the disease or wound is cured.

If we take some fish or animal flesh, we will see that many microorganisms are created on this surface. Where did they come from? Just from the meat itself.

In fact, meat is transformed into microorganisms, or in other words, meat and air create microorganisms. We will not see the microorganisms if the meat is frozen in the fridge. Even if we dip the flesh into liquid like vinegar, the microorganisms will not arise.

If we take air out of the atmosphere away from human beings or animals, we test it, we will not identify microorganisms in it. Microorganisms come from humans or animals. They live in our bodies and our skin. Microorganisms can not exist in the air for a long time.

When a man breathes he emits microorganisms into the air, and they can be found in his vicinity. Although we are not infected with a disease, microorganisms live in us. They live on our skin, in our respiratory tract, in our hair, etc.

If a man is sick he will let the microorganisms out of that disease. This living bodies will not live long in the air. They can not live on the skin of another man or animal for a long time.

The shape of the microorganisms is variable depending on the variation of the disease. Different diseases change metabolism over time, so microorganisms like their product change shape and other characteristics.

****

Medicine mentions the term called immunity. What is it in reality? Immunity means a person is resistant to diseases. His body is very healthy. The disease will not easily occupy his body. His body will fight the disease more than other bodies.

What is it that forces our body to fight? What is the factor that will trigger our body to fight? What are the elements of the body that are fighting the disease?

The coldness of weather can drive our body to fight. The body will respond with the help of its organs against such a situation caused by the cold. First, the nerves will react, and will stimulate body organs to counteract the new situation caused by the cold.

In such cases, stronger bodies can overcome the cold. In such cases, stronger bodies can stand against the cold, while the weak ones will fall. If we have ten people who come into contact with extreme cold, it can happen that only one man survives, and the other nine die. What happened in this situation? Why did this man survive, and the rest have died? His body was stronger and he was preparing for the cold.

When a flu occurs in some area, usually a few people die. Many will be affected, but most will survive the illness. We say that people who resist flu have good immunity. Their bodies are stronger and more capable of fighting the flu.

In the old age people are eating onions, garlic, hot peppers, etc. to cope with the disease. They used to take sliced lemon with sugar, or boiled red wine. This was a folk medicine that people have been using for centuries. Folk medicine drugs have served to keep the body resistant to various diseases, in fact to keep us from the flu.

Here we have come to the activity of our body, which must resist the various seizures that we can identify or not. Cold for example we can identify, but other different influences from the celestial space we can not.

In all these cases, our body will react, and will try to fight against a foreign attack. Medicine will help our body to win the fight. This fight we can call a sickness of cold or sickness of flu. In other words, we are fighting against the foreign factor that has attacked us.

The definition of illness is the struggle of our body against the external influence that has attacked us. External influences on us, or "Foreign Factors", can be many that will harm our body.

We can measure the high temperature of our body. We cough and sneeze on the nose, but in fact our body fights against the "Foreign Attack", which is hard to counteract. Imagine that arrows hit us, and our body fights with all its organs, muscles, nerves, the brain, etc. The aforementioned "Factor X" is the reason why our body fought against the "Foreign Attack" and produced a disease. During the fight, our organs will be damaged, such as the sluice, the nerves, etc., and we can win or overcome this fight.

According to these explanations, flu is the state of our body that fights against the "Foreign Attack". In these cases we can not see the enemy, but we know it is attacking us, and our body is fighting. This condition of our body is called cold or flu sickness. In the past, they used garlic, onions, boiled red wine, etc. to help the body overcome the disease.

The attack by the "Foreign Factor" has changed the condition of our body, that can be cured or collapsed, depending on the flu, the condition of the body, the medicine, etc. This condition of the body can be called "Resistant State". Our body fights with all its organs, such as the heart, blood, blood vessels, and so on against ilness.

Our body temperature will rise. We will feel tired, cough and sneeze from the nose. Our mucous will be swollen and red, but our body will fight. It will fight with all its capabilities and strength. In fact, the body will try to adapt to the new situation and to exist in it. The struggle will cause the diversity in our body, which will be shown in the composition of our blood.

How can medicine help us? It will calm our body. Our increased body temperature will fall. We will feel much better. Our body will relax. If our body fails to fight, our heart will stop, and we will die.

Here I explained the case of flu. Such a disease needs adequate medicine. It is also important in what condition the body or immunity is. Is our body strong and resistant to flu? Can our body withstand the fight with the flu?

What are the real remedies? How to find the right medicine to fight flu? We can use relaxing medicine, which will help the patient to lessen the pain, but it will not cure him.

In fact medicine is not that usefull. It only helps to reduce the temperature and calm the body. In the end, the patient will die. His organs have no power to fight this difficult situation. The man is weak and tired. We need strength to fight with the disease.

Vaccination does not help with flu-like illness. The flu is like a bullet which hit the body, and its organs struggle to survive. We can not protect the body against bullets producing the flu.

Influenza is actually the condition of the body. Our body was affected, and its condition changed. Body organs are fighting against the wound, i.e. disease.

Natural medicine for flu and cold were garlic, onions, some brandy, and red wine. People used to eat garlic and onions constantly as a flu prevention. Why was garlic an old flu medicine?

The body affected by the flu, means that our nerves are attacked. Nerves control the whole body and all its organs. If we take aspirins or a similar medicine, it will relax our body. Our nerves will relax, and the body will function better, but it will remain sick. We still have flu sickness. Our temperature was reduced by the help of aspirins. We can sleep and rest, but our body is still sick. Aspirins did not help our disease at all, but only to the body.

Here we have to differentiate body and illness as two separate elements. So if we help the body, we do not help the disease. With a help of aspirins, we live more pleasantly, sleep, get up from bed, but the disease does not change. The lack of intensity remains the same or increases depending on the stage of the disease.

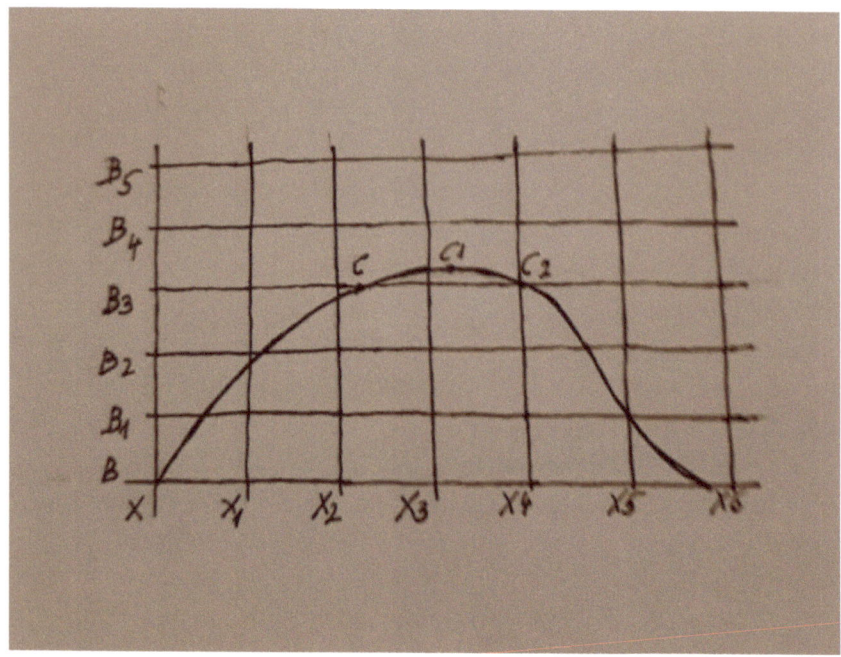

Curve of illness

B, B1, B2... Disease intensity values are shown on the scale "Y".

C, C1, C2... Extreme points of maximum disease value.

The disease started at the point "X". The values of its intensity are shown on the "y" axis in the points (B, B1, B2 ...). The time duration of the disease is shown on the "X" axis in the points (X1, X2, X3 ...). Points (C, C1, C2, C3) are maximum values of the strength of the disease, which can be read on the "Y" scale.

We can put every illness in a similar curve to reflect its value and intensity. This way we can predict all the details of disease on a graph like this.

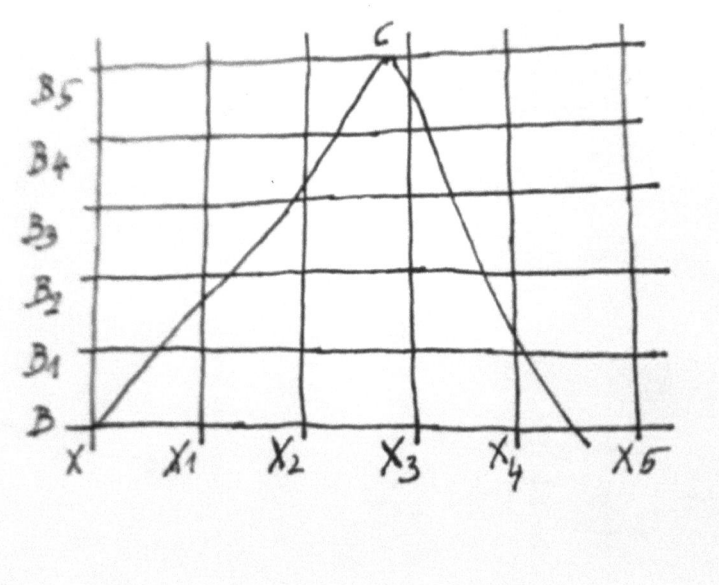

Curve of the disease with a sharp rise and fall.

B, B1, B2... Disease intensity values are shown on the scale "Y".

X, X1, X2... Time points.

C - Extreme point of maximum disease value.

On these sketches we can see two curves of disease. Both sketches have a rise and fall. Any illness can be shown this way. According to this, each disease in the curve has its start, rise and fall.

According to this many people will not survive the disease. They will most likely die in points (C), or before that. Medicine can use this graph to demonstrate the disease. Doctors can follow the disease on this map and add other elements like temperature, blood pressure, etc. They can predict disease development by drawing graphs of the future of the disease.

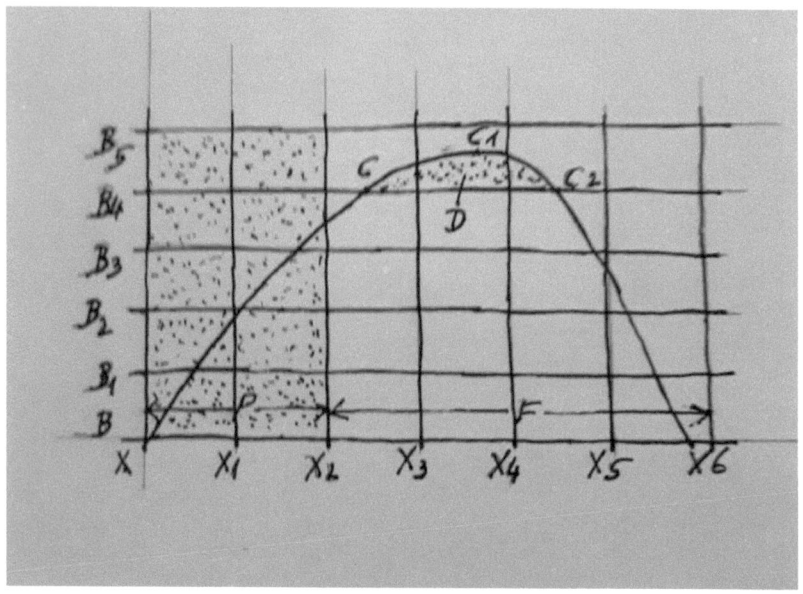

A future illness chart

B, B1, B2... Disease intensity values are shown on the scale "Y".

X, X1, X2... Time points.

C, C1, C2... Extreme points of maximum disease value.

D - Extreme area of illness.

P - Period from the past of the disease.

F - Period from the future of illness.

In the sketch we can see the past and the future of the disease. The doctor predicted and drew the future. Part of this sketch to point (P) at time point "X2" is the past of the disease. The value of (F) from time point "X2" to "X6" is the future of the disease.

In this way the doctor can see the future of the disease and apply the proper treatment to the patient. In shorter periods, the doctor can correct the graph and align it with real values.

The next medicine, which can be effective are antibiotics. Antibiotics should cure inflammation.

Let's say we have ten flu-affected patients. Nine of them will die and one will survive. Can we analyze this situation. Why did nine patients die and only one survived. They all took the same medicine and the doctors treated them equally.

Who were these people? One of them was an old man. The second was middle aged but very heavy. The third was a child etc. The man he survived was fifty years old, in good condition. He was neither thick nor skiny. He is an athlete, even though it is not a guarantee for survival. He could easily die.

The main point is that all of these people were in the flu zone. The condition of their bodies was not decisive. It is logical that stronger bodies will endure the disease better, but this is not a guarantee for survival.

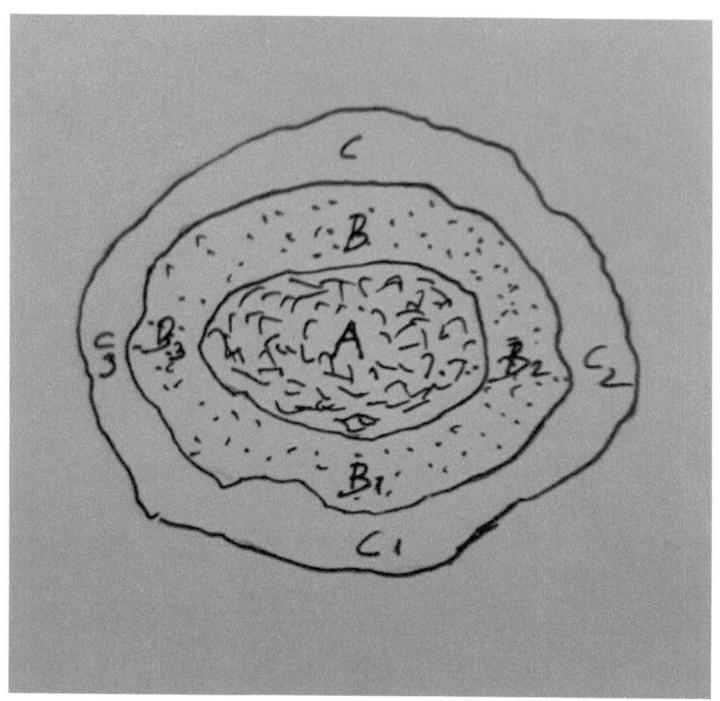

Flu Areas

A - The area of the strongest flu.
B - Less power flu range.
C - The weakest flu range.

In the picture we can see area "A", which is the zone of the strongest flu. "B" is the area of weaker flu activity. "C" is the area of the lowest flu.

In area "A", nine people out of ten may die. In the area of "B", it may happen that five people die of ten. "C" is the area of weak flu.

Here is the relationship between area and disease. In all these areas A, B and C there are flu symptoms, but the strength of the disease is different.

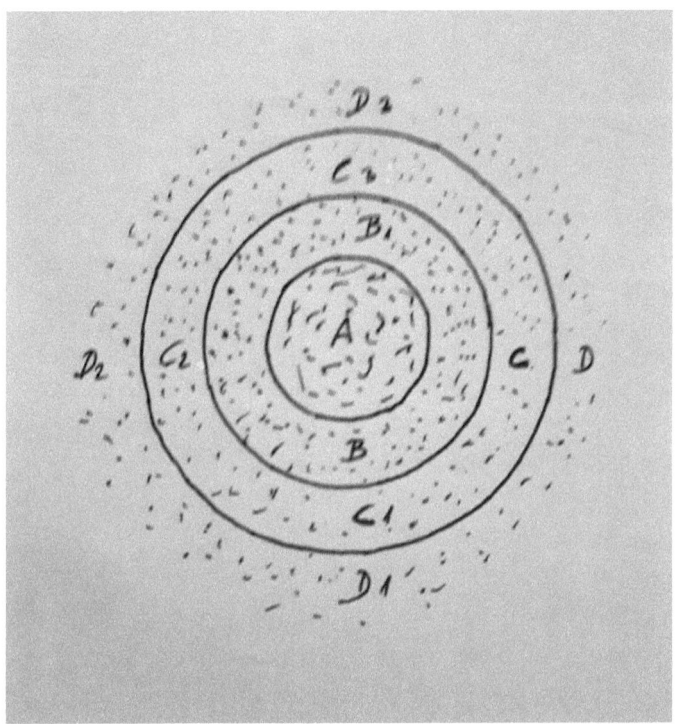

Areas of varying flu power

A - The main mass influenza area.
B , B1, B2... - Secondary range of flu.
C, C1, C2... - Light flu area.
D, D1, D2... - Free from flu area.

This sketch can be drawn whenever a flu occurs. Central area "A" will be in the area of major flu mortality. We can say that from every thousand people, one will die. Area "B" is a secondary zone of flu power. In this area, one in every ten thousand people will die. Area "C" is the third area of flu power. There one of hundreds of thousands will die there. No symptoms of flu were detected in area "D". As the flu moves from east to west, these circles can be visible on maps daily, weekly, etc.

****

Throughout history, mosquitoes were considered to be malaria transmitters. Recently, a "Zika" virus was detected in South America, which was transmitted by mosquitoes. In fact, it was reported that "Tiger" mosquito was transmitter of "Zika" virus.

"Tiger" mosquito has transmitted malaria, and now it has been transmitted "Zika" virus. According to medicine, mosquitoes flee from man to man and transmit malaria microorganisms. Thus, mosquitoes have viruses, or malaria microorganisms, and transmit it to many other people, perhaps thousands, tens of thousands, etc.

Where does the "Tiger" mosquito keep malaria viruses? Are they in the stomach or other part of the body? It actually looks like this. The mosquito bites a man who has malaria. According to science the mosquito should take malaria viruses from the sick man and transfer to healthy one.

My opinion is quite different from this. I believe that when the mosquito prick a man, it spurt its poison into him. This poison is actually a mosquito product. It is something like insect bites or snake.

The mosquito bite will cause the growth of microorganisms in our body, which will fight against the poison of mosquitoes. Newly discovered microorganisms are in fact malaria viruses.

Insects bite our bodies to reach our blood, or the liquid composition of our bodies. They actually feed on the blood components. Some other insects like bees will bite us if they are in danger of us.

Many insects bite our body just for food and one of them is a mosquito. Its whip and the poison, which it inject into our body creates malaria. All the microorganisms created after the bite was created by our body. Microorganisms of Malaria and Zika viruses are actually the products of our body.

The zika virus was created by the mutation of the Tiger mosquito. The Zika virus was created by the mutation of the Tiger mosquito, which has a new kind of sting. So this time the poison is different from the one that creates malaria. We have seen before that if we catch the cold, it will raise the microorganisms as viruses in our body. These microorganisms are the product of our body due to inflammation. Cold did not bring them, but they were created in our body. This is a good example of the formation of microorganisms in our body based on an external factor such as cold.

During the inflammation we can detect microorganisms in our body. They were created in our body due to "Factor X", i.e. because of such anomaly in it. In fact, something strange happened in our body, which did not work as before. Microorganisms will be rised.

The main control of our bodies are the nerves. Whatever they receive from the outside world, they transfer to our brain and organs. If they receive signals of cold, the nerves will transfer it. If we have a sense of coldness in all our bodies, the body will react immediately. The internal organs will react to fight against the cold. Because of this, the microorganisms will increase in our body. We can detect them in our blood if we analyze it.

Immediately after the body receives the signal of cold, it will enhance its ability to fight against it. The body will push blood faster. All other organs will use all their capabilities in this fight. After a while, the cold will catch our body. We will be able to identify microorganisms in the body, which is evidence that one is sick.

The same kind of illness will arise if the southern wind "Siroko" blows in Croatia. The weather is not cold but it is rainy and humid. Our nerves will indentify that our body has been attacked.

Similar types of disease will occur if the southern wind "Siroko" blows in Croatia. Then the weather is not cold but it is rainy and humid. Our nerves will find out that our body has been attacked. All microorganisms that occur in these cases come from our body. They are the product of the disease. If they are transferred to another man they will have no effect. Such microorganisms can not convey disease to another body. In fact, they are the result of our illness, which can not survive on another body.

I will compare microorganisms with some other creatures that inhabit our bodies. Fleas for example suck our blood, so our body is their natural habitat. The next being is the louse that live in our hair. It sucks our blood. Flies also shed our blood, but do not live on us.

Microorganisms of group "A" can be compared with these entities. Such microorganisms live on our body. Our body is their natural habitat. They live on our body, they multiply and eat us. Accordingly, they will injure our body.

For example, such microorganisms are those that cause gonorrhea. They live on our sex organs. During their life there, they damage such a habitat. We are getting hurt because it hurts us.

Microorganisms from our body will develop in the same area. They are actually the product of our body. The cause of their emergence is microorganisms of gonorrhea.

In this case, we can identify the microorganisms of gonorrhea and the opposite microorganisms of our body. Here the gonorrhea microorganisms are in fact "Factor X", which attacked our body. They raised the microorganisms of our body.

"Tiger" mosquito bite creates malaria disease. In this case, that stitch is "Factor X". The body will create the microorganisms that are the result of this sting.

Here we will see that different "Factor X" will cause the emergence of different microorganisms. This means that the cold will cause the formation of microorganisms from the cold. They have such characteristics that we can call them "Microbes from cold".

Every disease has its own microbes. They are actually the body's creation during the onset of such illness.

We know flu microorganisms etc. They are always the creations of our body.

In the case of "Zika V" caused by the "Tiger" mosquito, I understood the mutation of the same insect. Many mosquito bites are not dangerous at all. There are many mosquitoes in Europe but we do not have malaria. Pricks of these mosquitoes do not create malaria.

When I was a sailor I saw so many mosquitoes in Canada, but they were not dangerous. The greatest sting of mosquitoes I experienced in France in the Saint Louis du Rhone canal near the port of Fos, but they were not dangerous as well. These mosquitos were so cunning that they were tucked under the clothes and hurled me there. There mosquitos The mosquitoes hurt a member of the crew so much that he had to go to the doctor because of the high temperature.

In Rotterdam, the Netherlands has a lot of mosquitoes but they are not dangerous. Once in the area, malaria was known.

According to this, different mutations of mosquitoes will create a variety of infestation or sucking blood from humans and animals. Some bites are completely harmless, while others may hurt and create illnesses in us.

There is a mosquito in Africa that can cause Malaria. These mosquitoes are such a kind of mutation. For example, the mutation of the "Tiger" mosquito can be in Europe or somewhere else that it is not dangerous. Mutations of malaria mosquitoes in Africa create malaria sickness.

In South America, the mutation of the "Tiger" mosquito has created a "Zika" virus. In my opinion, "Zika" virus is the creation of a human body that was bitten by "Tiger" mosquito. Malaria viruses are also human body products. The buzz of a mosquito repellent creates the reaction of a human or animal body. Such a body reaction turns into a disease. In this case, "Factor X" is a sucker of the "Tiger" mosquito. The viruses found in the body are the product of the body. The body reaction, after the sting of mosquitoes, will produce viruses that are the result of this incident.

The intention of a mosquito is not to kill a man or an animal, but only to suck blood or body fluids. Such juices are their food. If we compare the mosquito bite and the bee bite, then we can see the difference. The bee bites a man because of its own defense. The tissue around the bite is swollen.

The mosquito comes to the human body for food. We feel his sting as painful and itching, which we can hardly bear.

Fleas, louses, and flies hurt our body for food. They suck the blood and juices of our body. We can snap all their stitches. Some bites are light and others are strong and painful. If they make the illness of our body we can call them „Factor X".

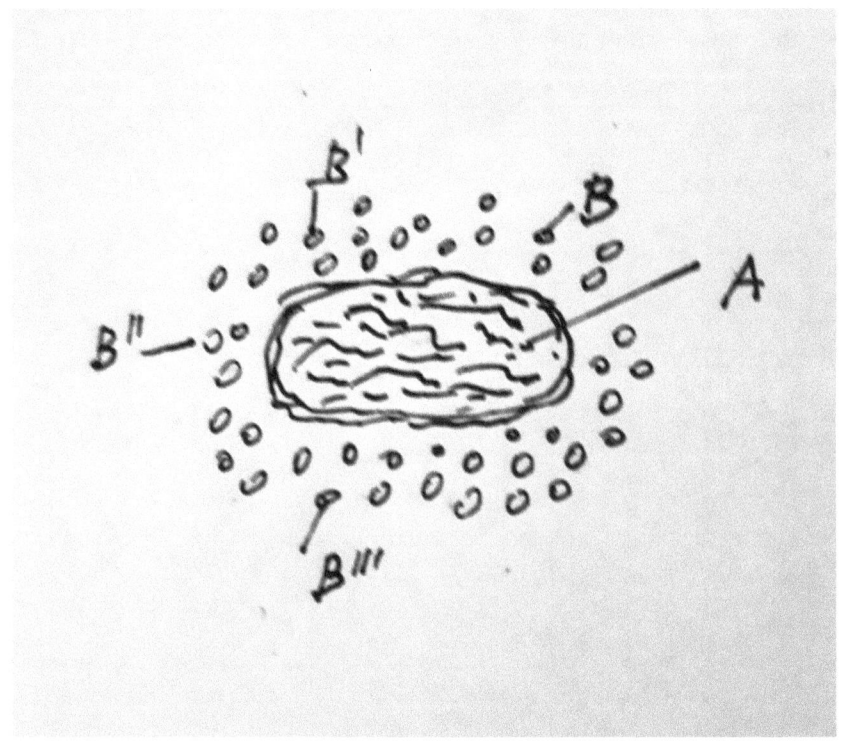

Creation of microorganisms

A – The wound at our body.
B, B', B''... Microorganisms formed on our body.

This is a very simple sketch of the wound (A) on our body and its surroundings filled with microorganisms (B, B ', B' ...).

I drew these microorganisms round. They could be formed in any other form. On this sketch they are formed as small round tiles.

Formation of microorganisms depends on "Factor X". Each new "Factor X" will form new microorganisms. This means that the shape and appearance of microorganisms depends on "Factor X".

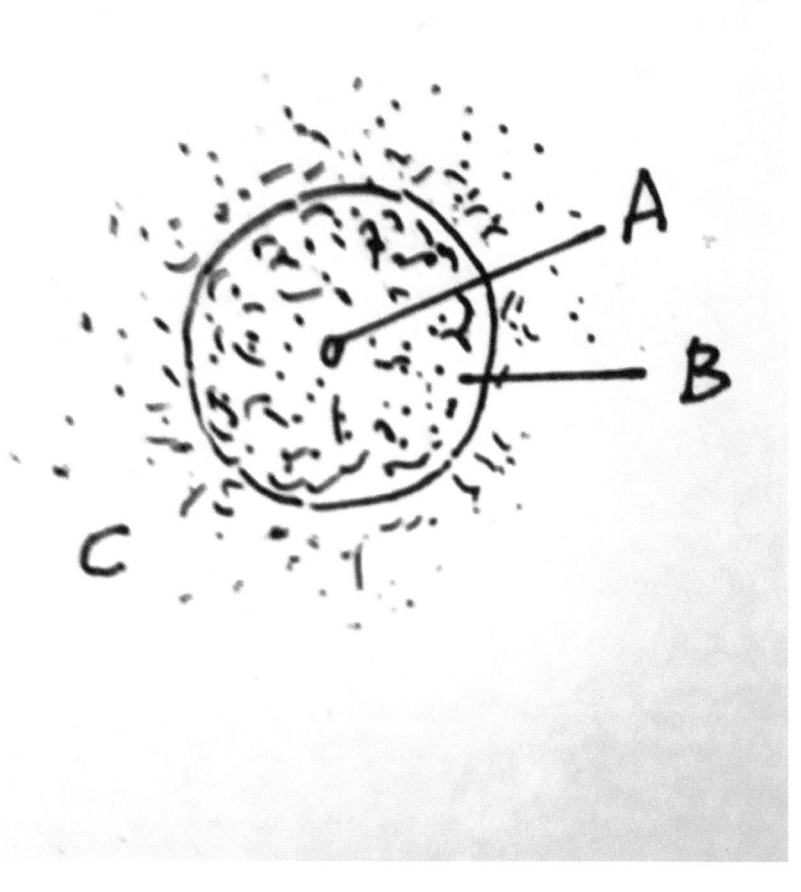

Moscito bight

A - Moscito bight.
B - The surrounding tissue, swollen.
C - The surrounding tissue, non swollen.

In this sketch we can see the sting of a mosquito (A). (B) is the surrounding tissue that is swollen. (C) the skin of the surrounding tissue, which is not swollen.

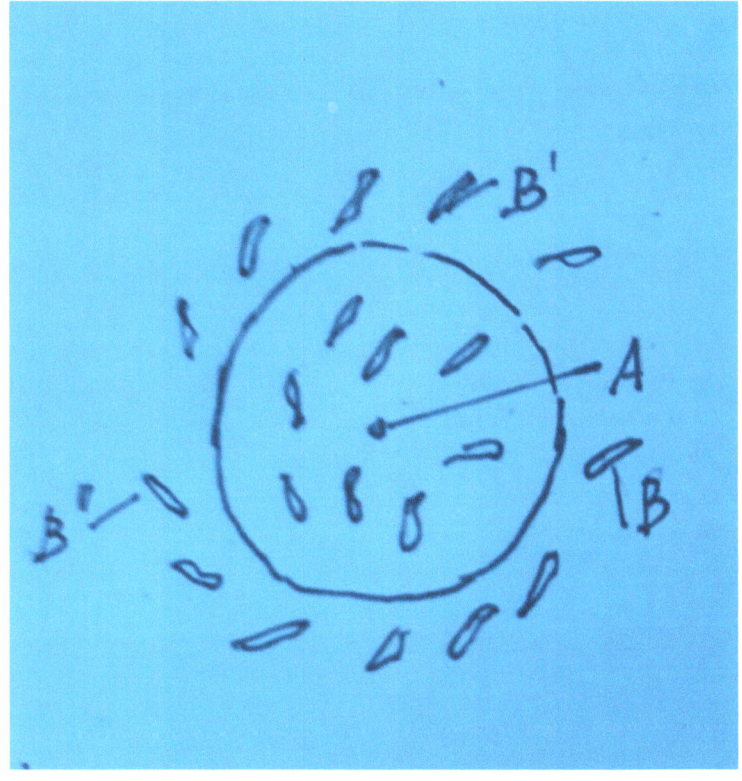

Demonstration of microorganisms after mosquito bite

A - Moscito bight.

B, B', B''... Microorganisms

In this case, B, B ', B'' are microorganisms that were created in our body during the disease, after the mosquito bite. They will enter our blood where we can analyze them.

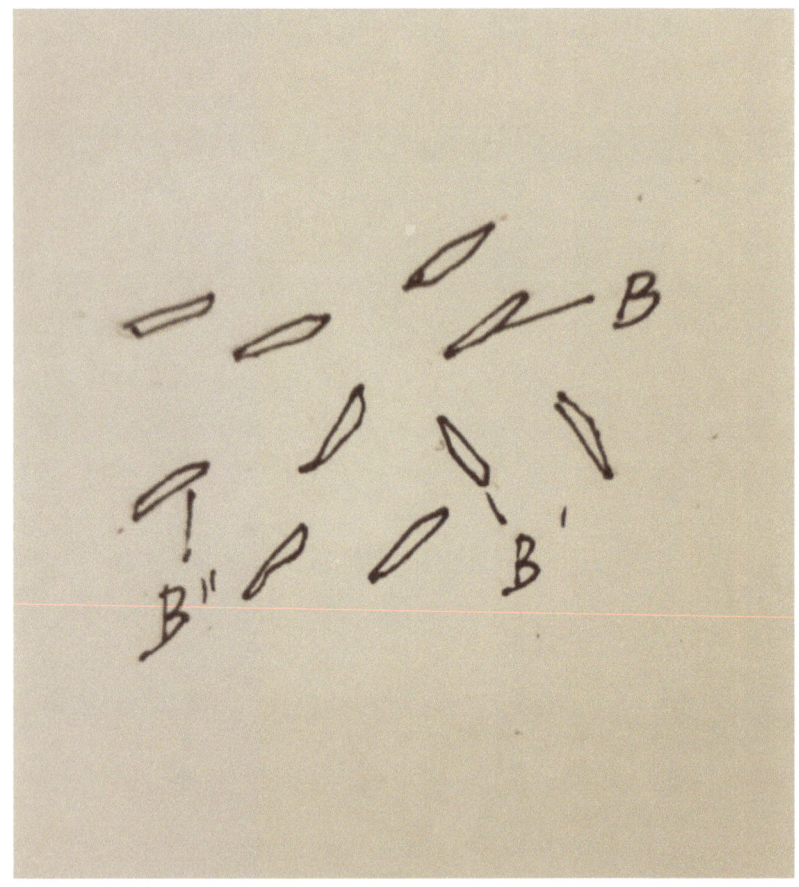

Demonstration of microorganisms form after mosquito bite.

In this sketch we can see the microorganisms that our body produces after the mosquito bite. In the case of malaria, microorganisms will have a certain appearance.

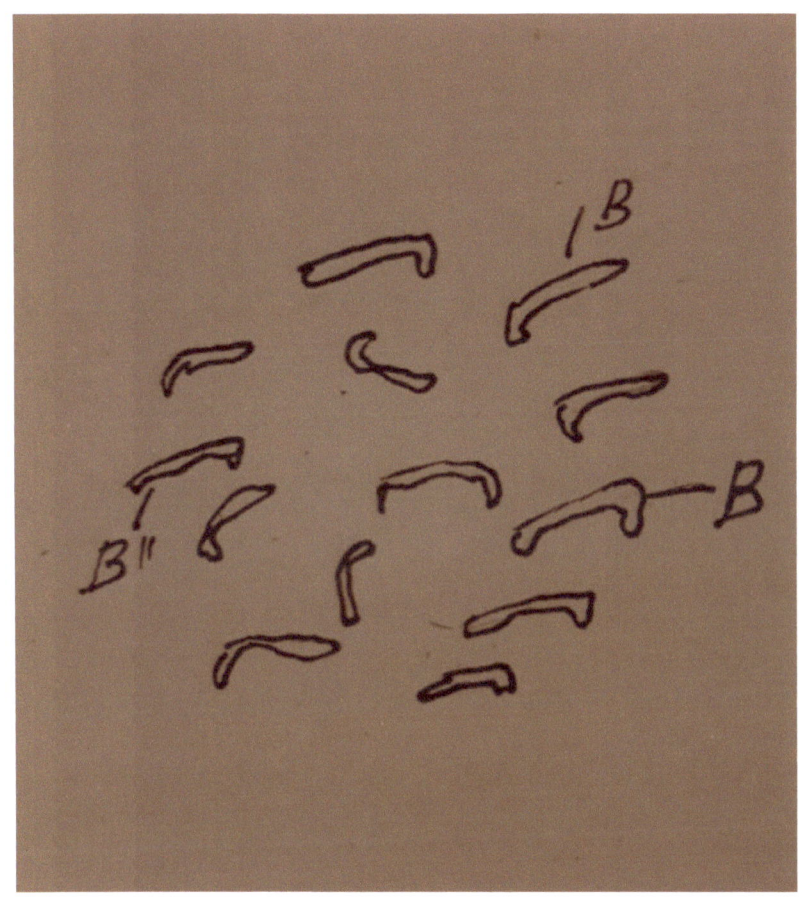

The appearance of microorganisms of diseases such as malaria, zika etc, caused by mosquito bites.

B, B', B'' - Forms of microorganisms of the disease caused by the mosquito bite.

We see the microorganisms (B, B', B'') on the sketch, caused by the mosquito bite. In the latter case, the microorganisms were flat and now they are slightly curved.

Any disease caused by mosquitoes will in turn cause various microorganisms. Different types of malaria or zika will create different forms of viruses.

****

If the fleas sow our body, it will not cause the disease. The site of the sting on our body will kick us, but it will pass. No illness will occur in our body.

Microorganisms such as viruses or bacteria are disease indicators. They can be perceived in the blood or somewhere else in the body, like a saliva, when the disease is dominated by our body.

The flea shoot is not so strong that it cause disease in our body. Likewise, fly sting will not create the disease, but the sting of mosquito is dangerous.

Here we can implement a "Geographic factor". Mosquito mutation is associated with geographic area. European mosquito is not dangerous. I heard that there once was malaria in the Netherlands. The mosquitoes are spread everywhere in the world. How can we know where malaria was in history?

According to the medicine, the Tiger mosquito is a transmitter from Malaria and Zika. If it sucks human blood, it can suck the viruses of some disease. Blood will come to its stomach. If blood gets into the mosquito's stomach, it will digest it.

If the mosquito conveys malaria, then other insects such as fleas, louses, flies etc. that bite our body would do the same. All these insects do not carry any disease. I believe that the bite of a mosquito creates the reaction of our body, and we are ill. In this case, "Factor X" mosquito sting.

Here we can see that our body is lying for some reason. Reasons like time, mosquitoes etc we can call here "Factor X". In this way we can find "Factor X" for any disease. What is "Factor X" of cancer? Can we find any cause for cancer.

Cancer is a disease that lives in our body. It destroys our body to form its own structure.

The structure of cancer in the human body

A - Cancer
B - Mass of the human body.

The cancer structure (A) is actually the mass that lives in our body (B). Its composition is actually a part of our body.

We can compare cancer with a skin ulcer. We will find that there are similarities in these diseases. Skin ulcers use the tissue of our body for their existence.

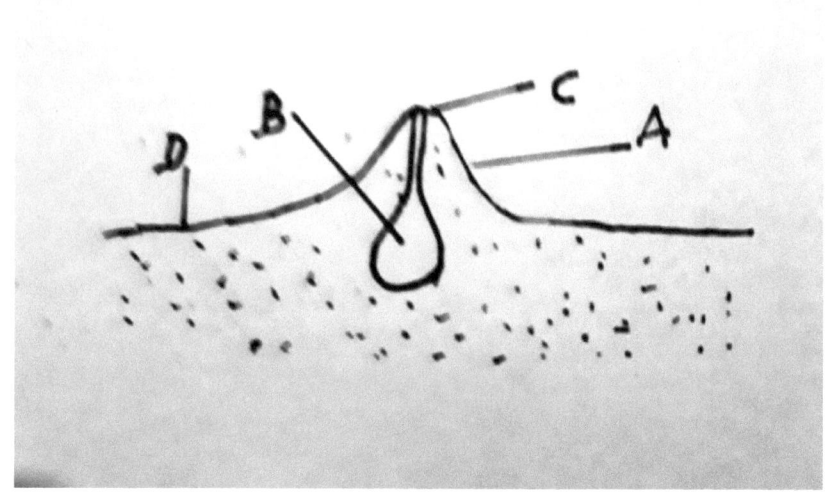

Skin ulcer structure

A - Bulging body of skin ulcer.
B - Ulcer with pouch of pus content.
C - Top of ulcer.
D - Man's skin with a tissue under it.

In the drawing we can see a pus bag (B) which is the main part of the ulcer. This is the major part of skin ulcer. (A) is swollen tissue under the skin. The ulcer tip (C) where the pus will go out. (D) is a skin tissue around the ulcer.

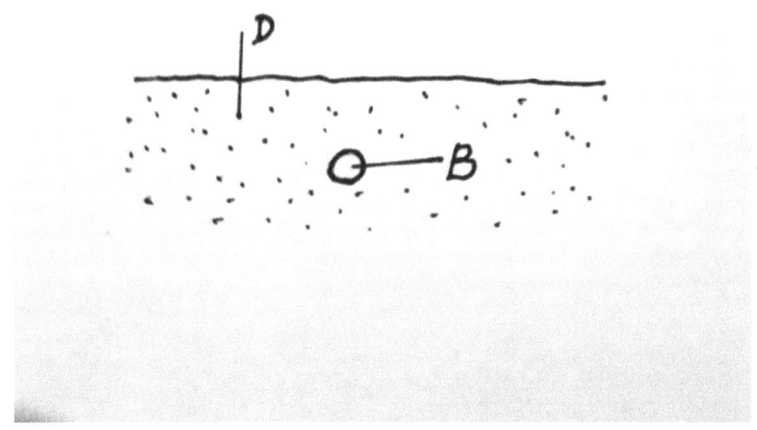

Early stage of skin ulcer

B - Begining of ulcer.
D - Skin tissue.

    In this sketch we can see the early stage of skin ulcer. (B) is the beginning of the ulcer. (D) is a human tissue around the start of the ulcer. The start of the ulcer (B) will find a way to form the channel and tip (C). The start of the ulcer (B) will find a way to form the channel and the top (C) on the sketch to discharge the pus.
    Here we see that an ulcer or any other disease uses our body to exist. Ulcer will actually eat our tissue for its existence. All ulcer activity will be fed by our body, which will exist using our body. In the end, the pus will be excised from the husk, and the ulcer will start to disappear.

We can associate this illness with a living creature. The ulcer is like a flea, which lives on our body. The fleas sucks our blood to live. It will die after a certain period of time.

In fact this comparison is not that adequate. The ulcer will be born and will live for a certain period of time. In the end it will disappear.

We can split the life of the ulcer into several phases. Here we will mention three periods of its life.

Cancer eats our tissue to form its structure. This is actually a new, living mass in our body. It creates its own tissue that came from ours.

If we look at the nipple, we will see that it has its own structure.

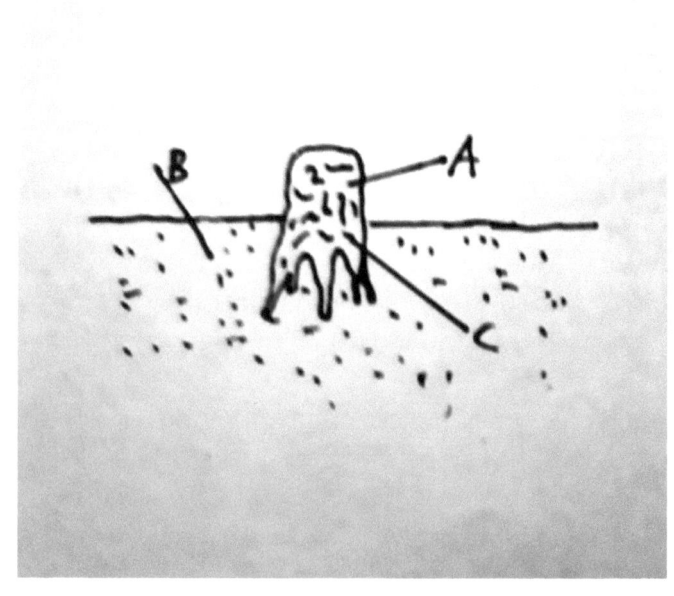

The structure of the nipple

A. Nipple
B. Man's tissue
C. Root

    The warts build their structure from our tissue. This is a very picturesque example. Here we can study the structure of some diseases like ulcer or cancer. These are the separated formations created on our body.
    The tree uses soil and water to build its structure. Mushrooms that live on tree trunks use it to build their own structure. How can we remove the mushrooms that live on the tree trunk? Sometimes their root is deeply embedded in the trunk. They bore tree trunks to spread their roots. If we get this kind of mushroom, one part of it will remain in the pit. After a while, a new mushroom will emerge from it.
    This is just a comparison that can help find the medicine for treating the disease. If we find a real poison that will not damage the tree, then we can remove the mushrooms.
    Warts are usually cured by surgery or rupture. If we break them, part of their structure will stay in our tissue. The second method is ignition. Today doctors remove the nipples with liquid nitrogen gas.
    The cancer will kill us. It will eat our vital organs and we will die. Cancer is like an animal that lives in our body.
    According to this our body serves for the existence of other species. Can we call the nipples a different kind of life, or are that part of our body. Are they actually a part of our tissue that has changed or deviated? Is this the other body that lives in us?
    In this way we have come back to the flea, which is another being, which lives on our body and it sucks our blood. In this case we can easily recognize another body.

In the case of a wart, we can hardly say it's the other being. We could say that this is part of our body, which has shifted to some other form. Something has affected our body to create nipples.

What was the reason that our tissue turned into nipples? We can compare it with the cold. If the weather is cold we can catch cold. In this case we can define our new disease. This has happened because of the cold weather. We cough and our nose is full of mucus.

Here we saw that we got the cold because of the cold weather. What was the reason we got warts and skin sores? We can not define it.

The main point here is to separate parasites and diseases that use our body for their own existence. Skin ulcers, nipples, lichens etc use our tissues for their own deviant structure. In this case something in nature has caused our body to function in this way, creating skin sores, nipples, and so on.

****

We can define the cold. I can say that I got cold for rain because I got wet. In this way I can find a reason for the cold. We can not explain every cold or flu.

We can say that a certain man has a cold because he has asthma. A person with asthma is sensitive to weather changes such as cold, rain, etc. Other people who live nearby did not get cold.

People with bronchitis are more likely to get colder than others. This means that our body creates a cold because it is not resistant enough. The composition of such tissue is different from other resistant tissues.

In the case of cancer, can we use the same comparison. Can we say that one man has cought cancer, since his body is sensitive, or it is inclined to it? Can we go this way, to compare the bodies, to define cancer.

We can say that one man will easily catch cold and the other will not. We can say that Peter always coughs or he sniffs his nose, but Mathew never catches the cold.

Can we use this kind of judgment for cancer in our bodies? Are some bodies and tissues sufficiently resistant to cancer? On the other hand, we can talk about "Cancer Season". "Cancer Season" in a particular country. Has cancer been known earlier? How will it be in the future? Will cancer exist in the near and far future?

How much I recall the skin sore season, I am not sure everyone had them. I remember well that I had them. Many of the diseases we have had as children have now disappeared, such as mumps, measles, etc. Seasons of these diseases have passed, while new ones are coming. In my childhood, children drank fish oil to be stronger and more resistant to disease.

Our body is so resistant to fight against weather and other sources of illness. How can we fight cancer? I believe the methods could be similar. If we manage to keep our body fit, it will help us fight the diseases. Such a body will create resistance to disease.

What is cancer in fact? First, we have to distinguish those diseases that are contagious, from those that are not.

If you have bronchitis or asthma, then there is a chance that you will get cancer. In this case, bronchitis and asthma are a cause for you to suffer from cancer. These causes and specifics of the body lead us to cancer. In this case, the bronchitis is "Factor A" of cancer. In other words, bronchitis or asthma have caused cancer to somebody.

Good and strong body can easily resist bronchitis and asthma. When the body becomes weak bronchitis and asthma will overwhelm it. When asthma overcomes the immune system, cancer will take the body.

Cancer can be the product of some diseases. In fact, bronchitis and asthma are the struggle or resistance of the body against external influences. In this way, the body fights against external influences, creating bronchitis and asthma.

Therefore, many diseases are actually the consequences of fighting bodies against foreign influences. "Factor A" is actually a disease created by such a struggle. Here I do not include illnesses caused by contagion.

The cancer will occupy our body, because we can not fight more. In fact, the body is so created that it can withstand foreign influences.

Cigarette smokers are considered to be lung cancer agents. I would not accept it and I consider that if the lungs are not sick, smoking will not hurt them. In fact cigar smoke only shows that the lungs are sick.

The lung cancer is biologically embedded in the lungs. In fact, it is a diseased tissue, which has transformed the lung tissue. Smoking cigarettes will only detect it.

Here we have found that cancer is actually a disease, which lives in the tissue of our body. The cancer takes away the healthy tissue of our organ and transforms it into "Cancer Tissue". Cancer transforms our organs into its newborn diseased tissue.

There were skin sores in my childhood. I remember that I had a few such ulcers, but some of the schoolchildren did not have them. This means that all the students could not get them. Each organism is different. Only some of us had been attacked by skin ulcers because our bodies were suitable for them.

Skin ulcer uses our body to build its own mass. Skin ulcers have disappeared before so many years. We now have a cancer season. Now the astrological conditions are suitable for cancer.

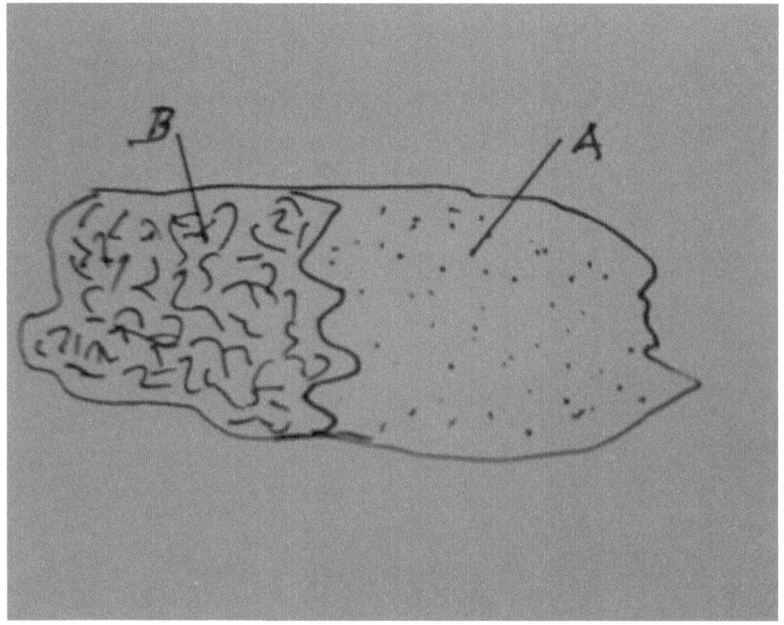

Part of the cancer-induced organ.

A. Healthy part of the organ.
B. Part of the cancer-induced organ.

In the sketch we can see part of the organs suffering from cancer "B", and the healthy part of the organ "A".Cancer has transformed the tissue of our organ into the "B" part or the diseased zone. Cancer uses healthy organ tissue "A" for its existence. The composition of healthy tissue will become zone "B".

We could say that zone "B" eats zone "A", or zone "B" occupies zone "A" with its illness. In any case zone "A" is transformed into zone "B" under pressure of cancer existence.

Here we have two existents, the existence of "B" as cancer and the existence of "A" as healthy tissue. In other words, existence of "B", or cancer, strives to overwhelm the existence of "A" or healthy tissue. Cancer often overwhelms this body battle.

To make this clearer, try to compare the ulcer and cancer again. The skin ulcer used our tissue to build its own structure. It could damage some of our organs, like blood vessels. In fact, it was not so dangerous. It had a width structure of two to three centimeters. Its depth was about half a centimeter.

According to this we have a few influences on our body. One of them is the climatic influence that can create or strengthen some diseases. The next is the celestial influence, that is, the position of Earth in Celestial Space. Such an effect will create mutations or changes in the illness. That is why the former skin ulcer disappeared. At the same time, today's cancer has emerged.

I do not want to link these two diseases. Here I list them as an example of disease dependence on the position of Earth in Celestial Space. In reality, all diseases are related only as they change their structure depending on the above-mentioned influences.

Climatic conditions will create mutations of our diseases. Let's say this year colds, rheumatisms and similar illnesses were not as severe as last year. In this way, climatic conditions and changes in the position of Earth in Celestial Space are manifested.

I consider climate impacts less meaningful than Celestial changes. Celestial changes are primary in relation to climate and similar changes, which are secondary. This means, for example, that last year's rheumatism was stronger and more painful than this year.

The effects caused by the Earth's position in the Celestial Space will change illnesses such as, for example, skin ulcers will disappear, cancer will appear. Who are we actually? Does the precession of Earth in Celestial Space form us? Do they affect our body?

According to this outline, we are like clay that various celestial influences form towards the relationship of Earth and Celestial Space. The shape of our body and all diseases are changeable in relation to the precessions of Earth in Celestial Space.

In ancient history, the Earth has created precessions in the celestial space that shaped us. Earth precessions affect us and all the living world on earth.

Here we can again compare skin ulcer and cancer. Ulcer was a disease during a certain precession period of earth.

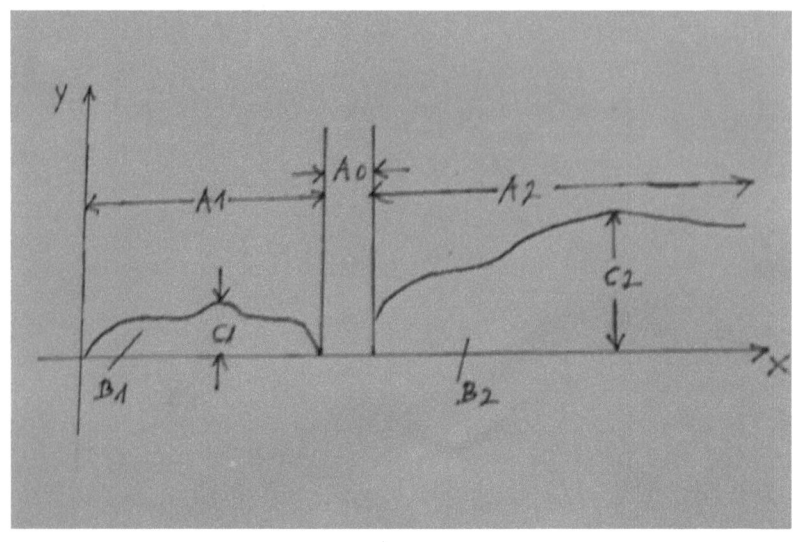

Period of illness

A1 - Skin ulcer period
B1 - Curve of skin ulcer disease
A0 - Zero period
A2 - Period of cancer
B2 - Cancer Disease Curve
C1 - The greatest strength of ulcer disease
C2 - The greatest strength of cancer disease

In this graph, we can compare the disease of ulcer and cancer. In addition to these two diseases, at the same time there were other diseases that are not shown in this drawing.

The skin ulcer's disease had its (A1) period, which can be labeled with age as 10, 20 years, etc. Period (A0) is the time interval between periods of ulcer and cancer, e.g. 20 or 30 years. I can not say that there was a zero period between skin ulcer and cancer, but I mention it because of the illustration.

(B1) is a graph of life of the ulcer. (B2) is a graph of cancer life. This sketch I just drew for illustration.

In this sketch, we see the ulcer's greatest strength (C1), which means that the ulcer was then the strongest, in fact the most widespread in a certain area.

(C2) shows the strongest cancer power, meaning that at this point, cancer covered the broadest area. Later cancer fell to the same level.

During the onset of ulcer, the drug was called a black ointment. We used it to make the ulcer mature earlier. Another good medicine was a leaf of a plant, with which we covered the ulcer and adjusted the bandage over it. We then waited for the ulcer to mature and soften.

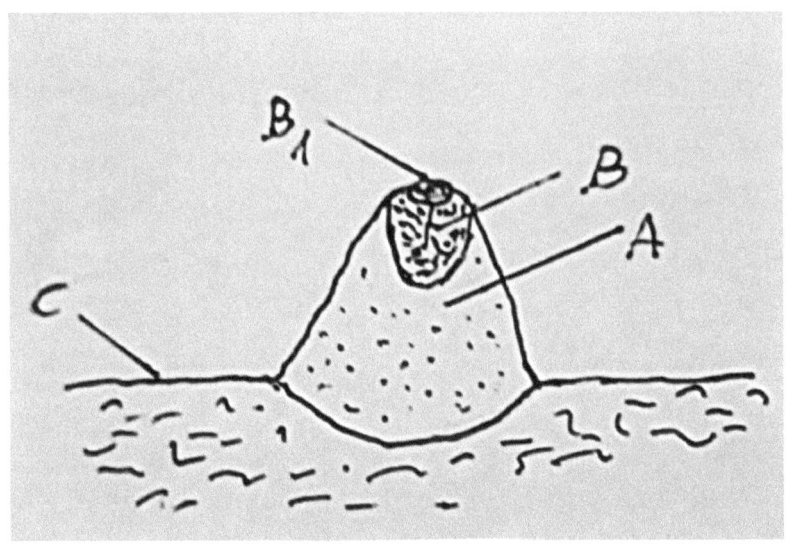

Mature ulcer

A - The body of the ulcer
B - The pit with the secret

B1 - The pit opening
C - Human skin

In this sketch we can see the mature ulcer. When the secretion is shed, the ulcer will end, and after some time remain the scar.

Today there are acne that looks like an ulcer. They are expressed as small wrinkles on the skin.

We live in a cancer period. First we have to understand what cancer is. How to cure it?

I have made a mathematical diagram for easy understanding of the disease. This diagram is based on the existing planimetric diagram.

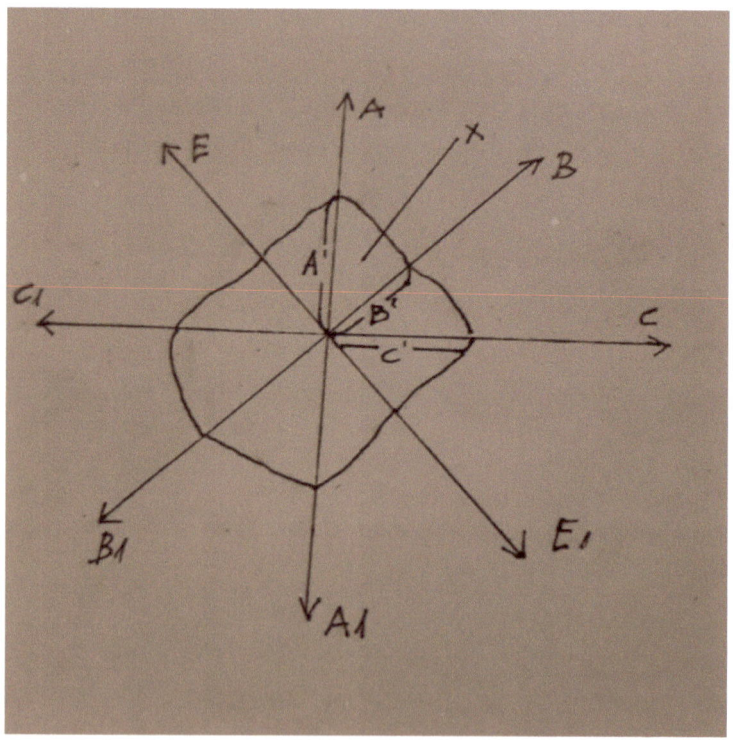

Clip of surface from cancer body

X - Surface of the body of a cancer clip
A – Temperature
B - Blood pressure
A', B', C' - Parameters of cancer medical values

In this sketch we can see many elements of the disease that can be labeled in the diagram as temperature, blood pressure, etc.

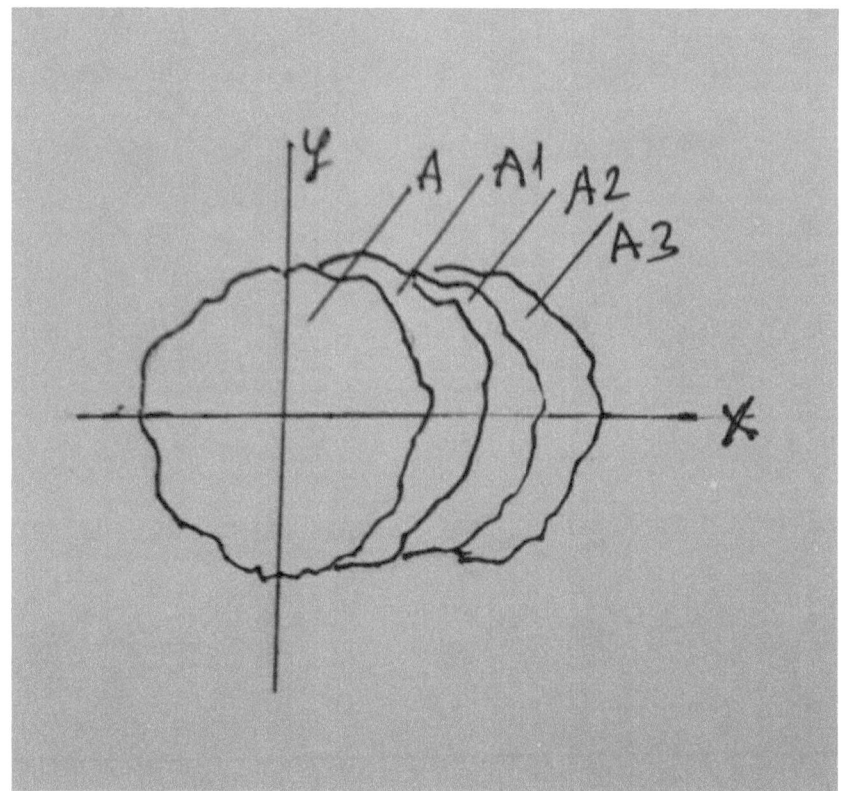

Body of cancer at x - y axes

The areas of the cancer body are separated by time periods of 1, 2, 12 hr or 1 day etc.

In this way, we have the cancer elements that we need. We can easily study them on a computer.

We can test different medicine and healing to see how the body will react to them. For example, if we give the patient lemon twice a day, which is the old medicine, we will see how the graph will look, whether the graph will be the same or different.

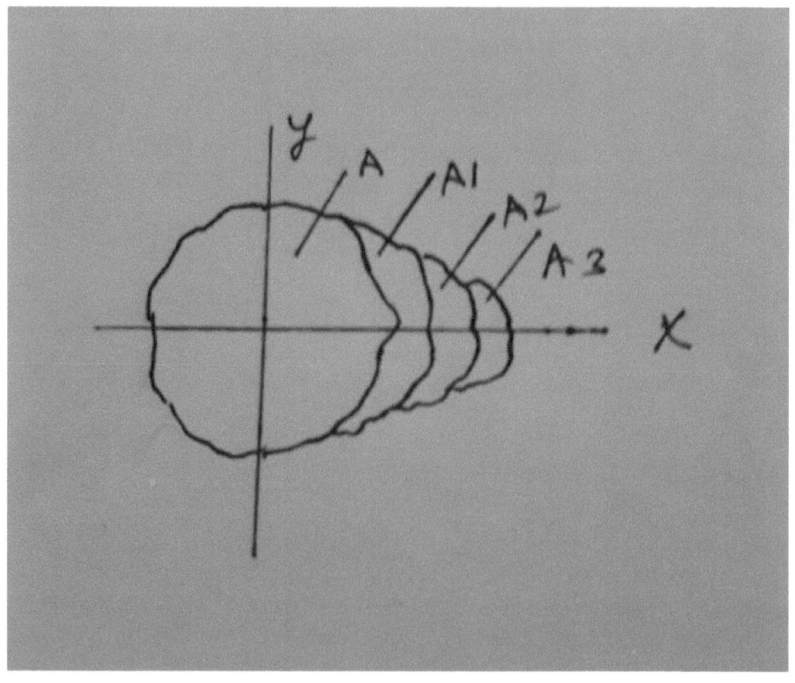

The cancer body after a certain treatment.

A, A1, A2, A3 - The surfaces of the cancer body.

Here are clips of the body surfaces of those elements that are diminished by the action of drugs. If we give the patient a certain medicine then we can follow that on the graph. A sketch chart shows that the drug is successful and the patient is recovering. Contrary to this graph would be the same as before, if the drug did not act to treat the patient.

In this way we can choose a cure that will heal the patient. On the graph we can follow the effect of any drug.

We can try natural remedies like garlic, lemon, honey, etc. Each of these drugs strengthens our immunity. This is the basis of treatment, as in a case of ulcer, when we used a black ointment, or leaf from the plant, to make the ulcer maturely.

How we treat cancer depends on us. Any drug we use will show the effect on the graph. Even garlic treatment will show the effect on the graph.

****

In addition to meteorological effects, our body is subject to celestial influences. Earth precessions create celestial influences on us and the rest of the living world. Just as Earth is set in a certain period of its precession, so will we feel in that period. Our body will resist influences, in relation to our immunity.

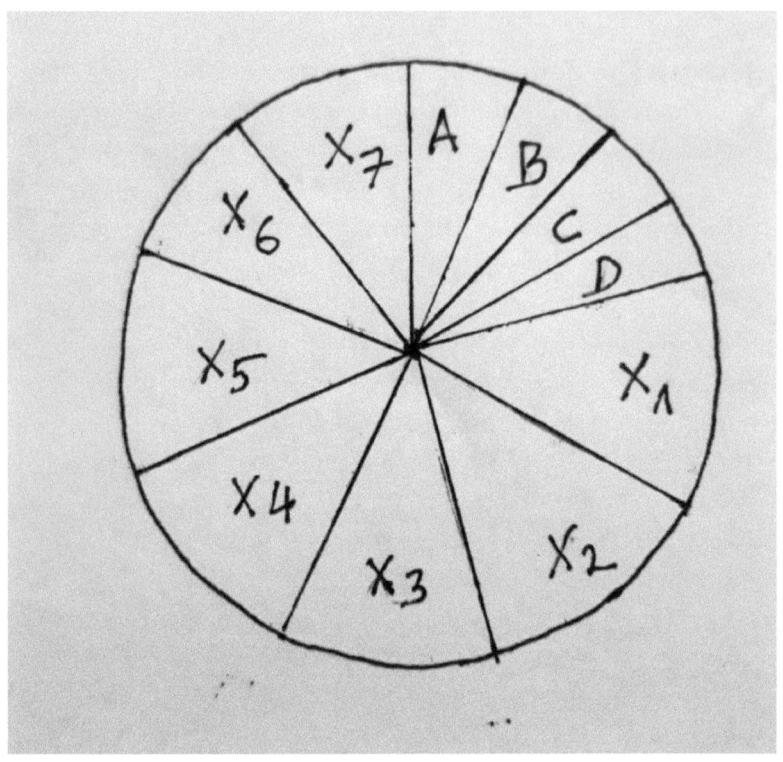

Impacts on the human body.

- A. Influence of solar radiation.
- B. The effect of cold.
- C. Impact of plants.
- D. Seasonal influences.
- X1....X7. Celestial impact.

    Can we identify any of the impacts (X)? What is X1, X2, X3 for example ... I believe we will be able to find that in the future. We can determine what diseases are caused by Earth effects, such as because of the cold, and which are due to celestial influences.

What exactly is the celestial effect? This is all that changes the elements of survival in a particular area. If Earth moved to another position, certain elements of survival were gone. Because of this, diseases have different composition depending on the precession of Earth.

The celestial effect can be divided into many elements, which is actually a set of different influences.

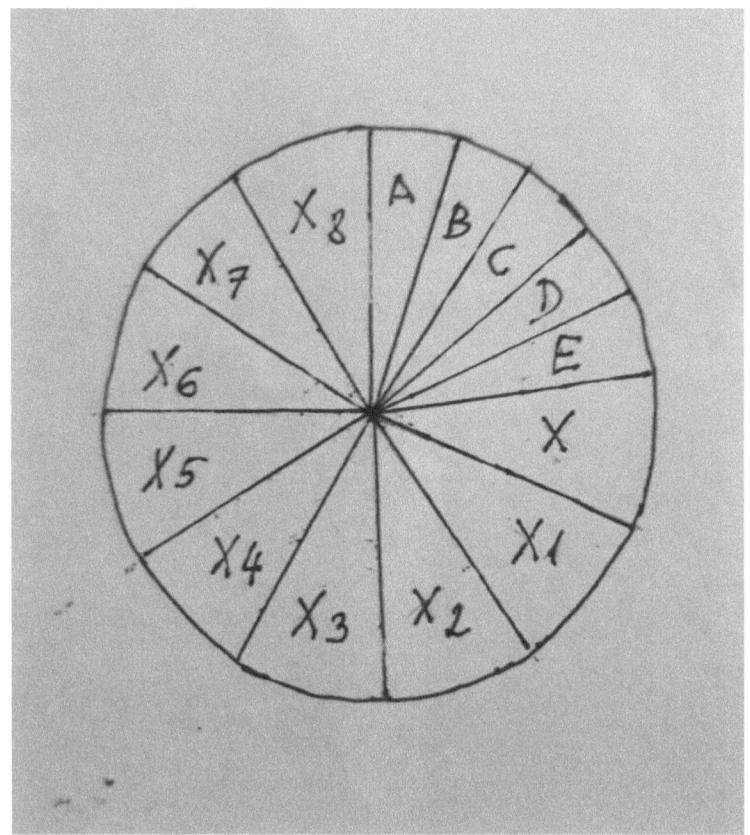

Celestial influences.

A. The influence of sun.

B. The influence of moon.
C. The impact of the planets.
D. The influence of stars.
E. Influence of general celestial radiation.
X – X8 Celestial indefinite impact.

Here we can designate some celestial emissions, such as radiation of sun, moon, etc. The X - X8 emissions are unknown effects to us, which actually exist.

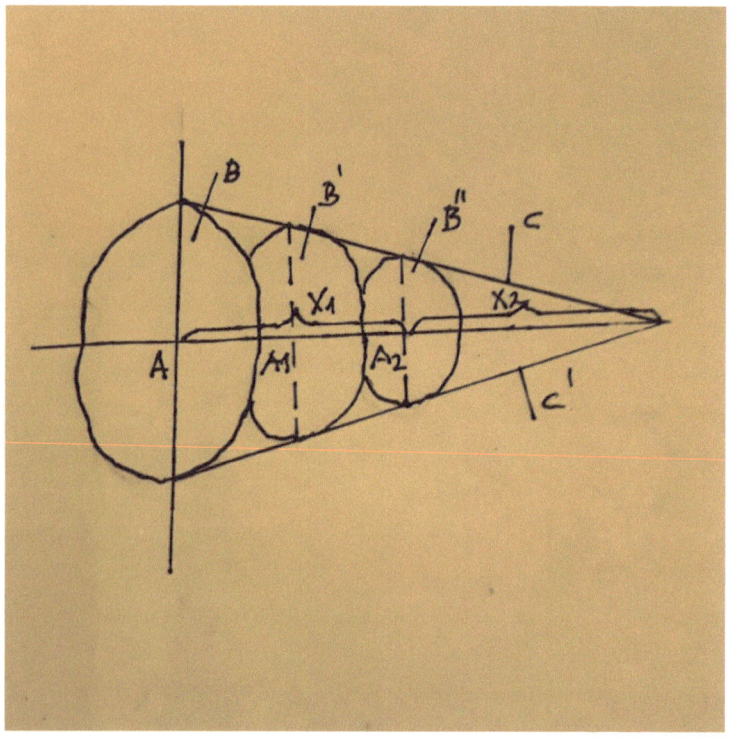

Test of medicine.
A, A1, A2... Test points.
B, B', B''... Snippets of cancer body surfaces.
C, C1 - Body lines.

X1 - The test line.
X2 - The line of the future.

In the drawing we can see the body of the disease. B, B ', B "are the surfaces of the body of the disease. They are timed as every hour, two hours, etc. A, A1, A2 are test time points. C, C1 are the outer lines of the disease body.

If we administer the medication to the patient for a specified period, then we say that every two hours we will take values on a section of the body of the disease. In fact, this drawing shows us sections of the body of the disease. According to this drawing, attributed medicine is effective. This is a look into the future of disease behavior.

We can say that point (A) was at 6.00 in the morning. Point (A1) was at 08.00 and point (A2) was at 10.00.

Based on three tests in the (X1) period, we can assume the development of the disease body in the (X2) period, which will continue in the future. This way we can study the life of the disease in the past and in the future.

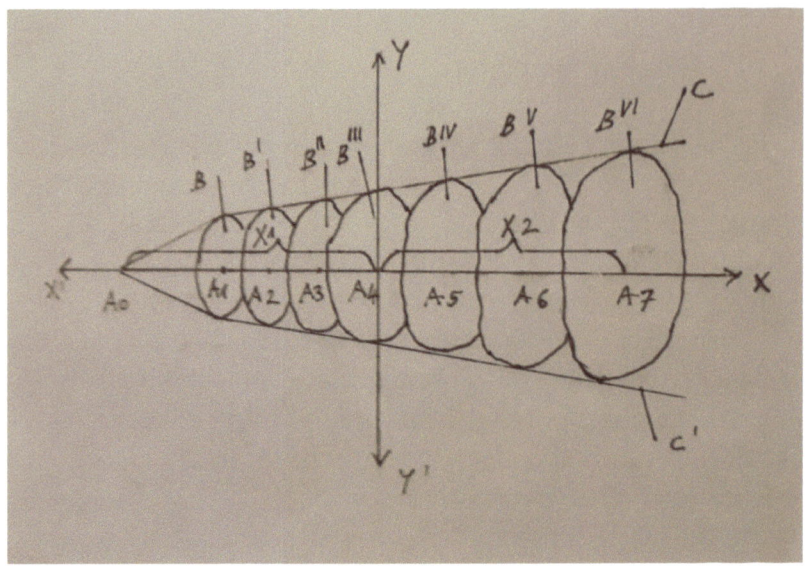

A life of illness.

A0, A1, A2... Points of the disease life period.
A1', A2', A3'... Time periods between plots from the body of the disease.
B, B', B"... Surfaces from the body of the disease.
C, C' External lines from the body of the disease.
A0, A1, A2... The points of the disease body where we take tests for periods of time.

Here we follow the disease from its beginning at point (A0) in the past to point (A7) in the future. The drawing shows that the disease is increasing even though we were giving medicine to the patient. From point (A4) to (A7), the calculations are in comparison with the past of disease.

We can draw such a graph in periods of ten minutes, one hour, two hours, etc. The values will be indicated on the graph plots.

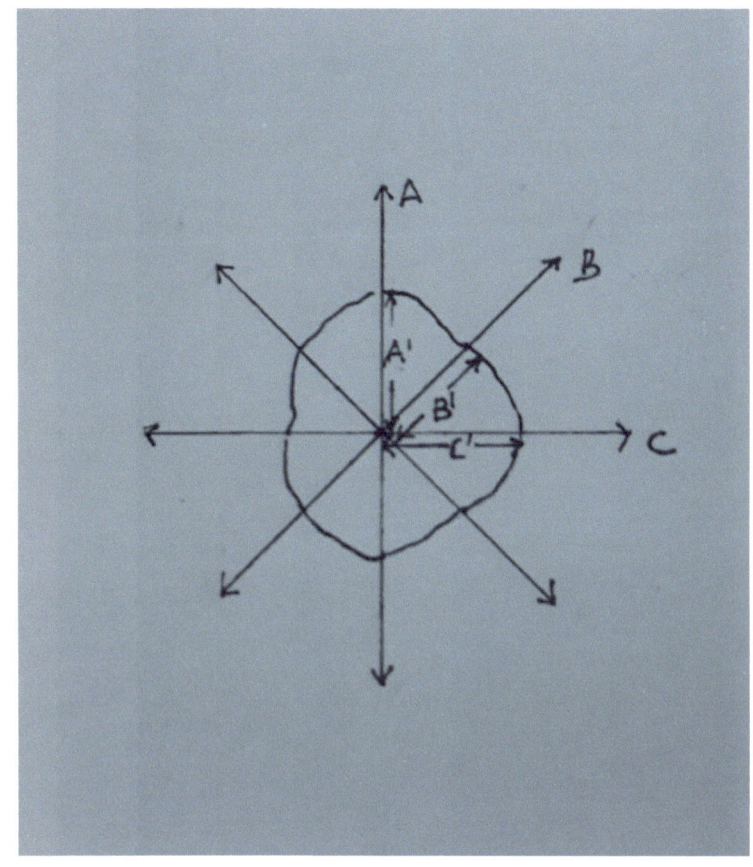

Area of body condition values, during illness.

A - Body temperature.
B - Blood pressure value.
C - Pulse height.

Based on this sketch, we can denote the values of different parameters of our body during illness. The disease graph can show many parameters such as mental state, mood, desire to eat, etc.

There was once a disease of ulcer, which disappeared. There are still acne or puffs, which resemble ulcer, but they are much smaller. Which caused the ulcer to become as small as acne or puffy?

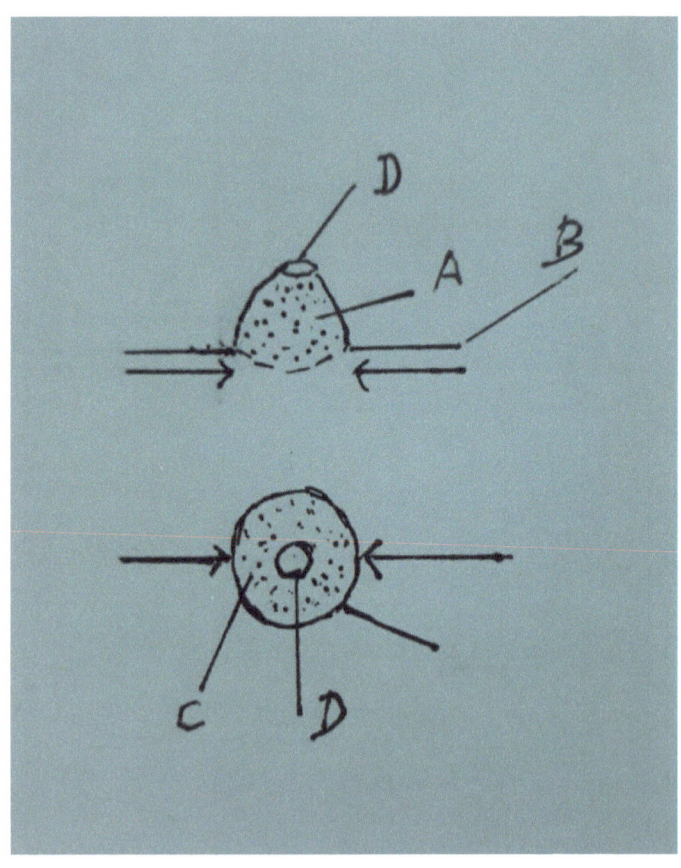

Acne

A - Acne
B - Human skin
C - Diameter of acne
D - Top of acne

    Acne is similar to skin ulcer, but it is small. The diameter of the acne is at most 5 to 6 millimeters. Its height can be 2 to 3 mm. It has a tip (D) where the secretion will come out. The skin ulcer had the same structure, but its diameter was 3 to 4 cm, its height was about 1 to 1.5 cm.
    Astrological science is preserved today in churches, as a Catholic, Orthodox and other religions. According to my book, Formation Of Planet, Earth and Planets do not revolve around the sun. They have certain precession paths. Astrology calculates the relationship between earth and other celestial bodies.
    I will use an example here that is not realistic. I actually made it up. "This year, over a period of time, Venus will be crowned with three stars called so-and-so." During this two month period, the flu will be very strong. People will suffer from throat diseases, etc.
    Here we can mention the astrologers' predictions, which predict that during the period from January 15 to March 15, some diseases such as influenza and throat disease will intensify. These diseases will be much more severe than last year, even a few last years.
    This is where we have touched prediction, or looking to the future. Meteorologists explore the future in their science. Medicine does the same. Doctors can determine the future of the disease and say when the patient will recover, or the disease will worsen.
    I want to mention here that colds, flu, sore throat, bronchitis, asthma and pneumonia have similar manifestations of the disease. These diseases produce secretions in which

bacteria and viruses are visible. I consider viruses and bacteria to be the product of a diseased body. Therefore, these microorganisms cannot live on another body, no matter what medicine says.

Here we can investigate how certain diseases arose. Are individual diseases caused by changes in weather, or cold, rain, etc. What diseases are caused by celestial influences?

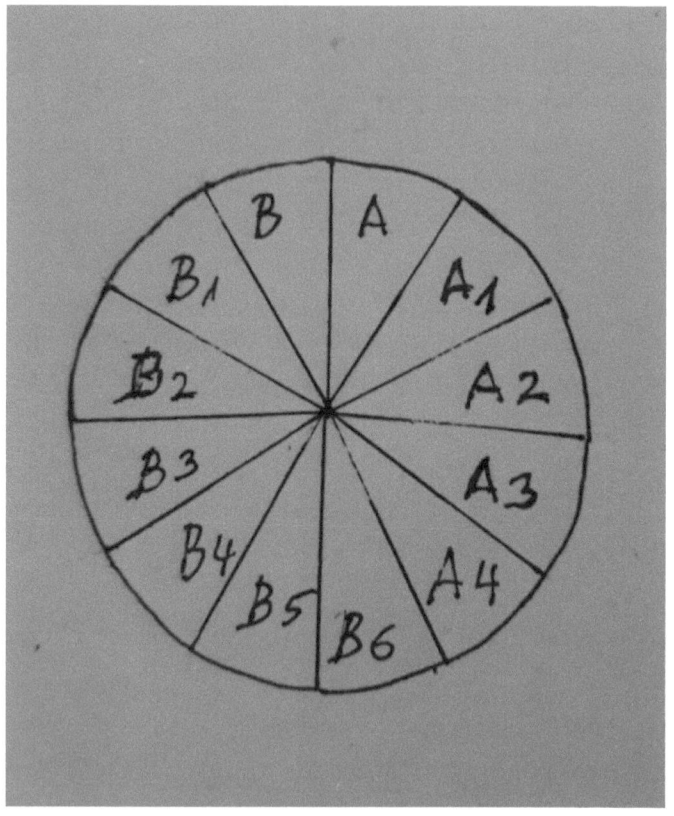

Disease distribution.

A, A1, A2... Diseases caused by changing weather.

B, B1, B2... Diseases resulting from celestial influences.

On the sketch, we can see all areas of diseases caused by meteorological impacts, such as (A, A1, A2 ...). These illnesses will develop due to cold, rain and the like. In case we get into the cold sea, it will be very easy to catch a cold, that is, we will become ill.

Group B, B1, B2 ... are diseases caused by celestial influences. They can occur in certain seasons such as winter, spring, etc.

The human or animal body can withstand all these influences of the seasons and the change of position of the Earth towards the celestial bodies. Earth moving will cause disease mutations, their severity and the danger to human and animal bodies. Because of these effects, the disease will change shape, severity, etc.

Here we can again compare skin ulcer with acne, so that we can compare the variability of the shape and size of the disease due to Celestial impact, which is variable and caused by the change in the position of the earth in Celestial space.

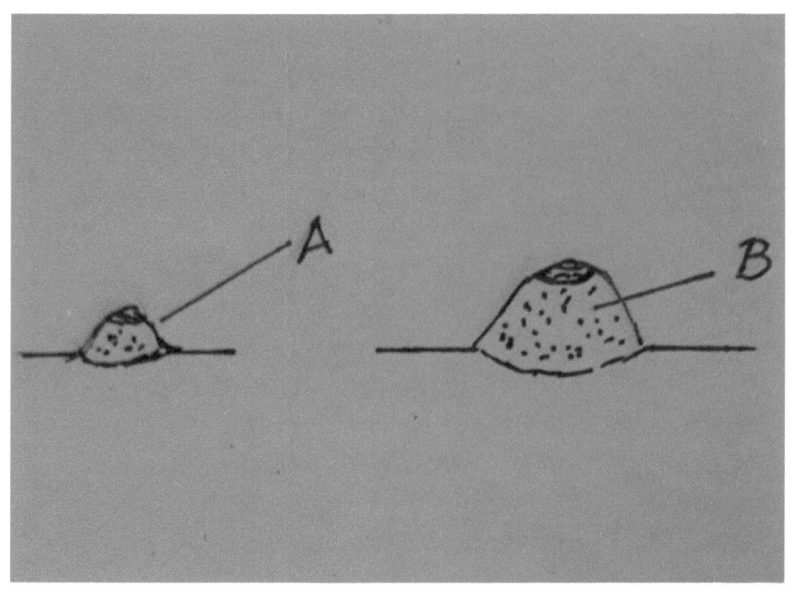

Comparison of Acne and Skin Ulcer.

A - Acne
B - Skin ulcer

On the sketch we can see the difference in the sizes of acne (A) and ulcer (B). The size of the ulcer is 3 - 5 cm, while the size of the acne is 3 - 5 mm. The variability of the disease in size and shape or its mutation is visible here. A particular disease will change in size and intensity, ie the danger to humans or animals. We can call this process "Disease Mutation". In fact, all diseases are permanently mutated, whether they are enlarged or diminished. The mutation of the disease depends on the variability of the earth's position in the celestial space.

In fact, the diseases do not exist by themselves. Under celestial influence our body will create certain diseases. We can call them "Celestial Diseases". These diseases are different from "Meteo Diseases", which are caused by "Meteo Effect".

Based on these explanations, we can study Cancer. What did cancer look like fifty, one hundred or five hundred years ago? Was it dangerous? How and why did it mutate into new forms?

****

Ulcer was usually below 5 cm in diameter including its swollen tissue. It was less than a centimeter deep.

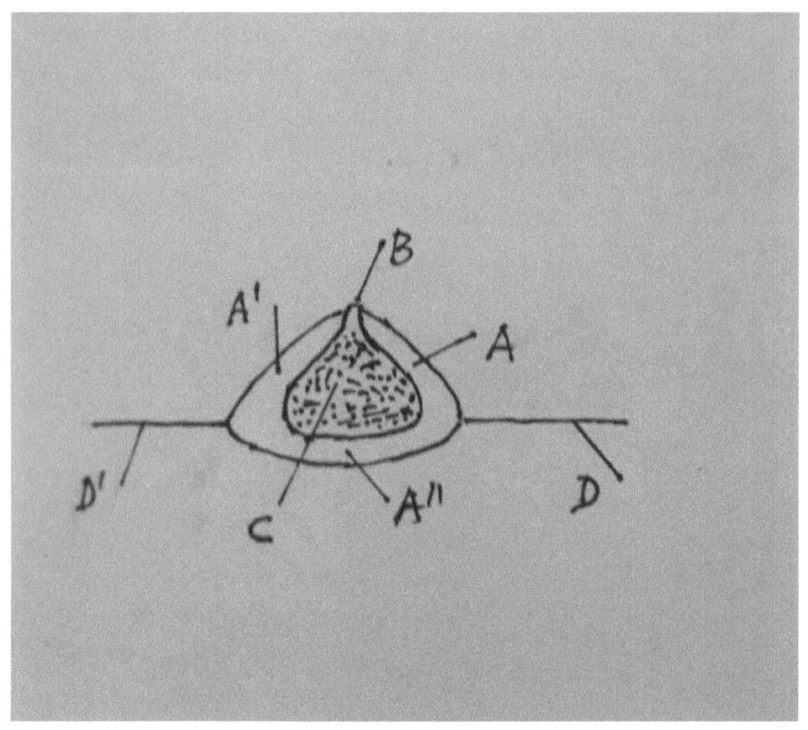

Skin ulcer

    A - Swollen tissue.
    B - The top of the opening on the ulcer.
    C - A purulent bag.
    D – Skin

    In this drawing we can see the purulent bag (C), which is dangerous for us. A, A ', A "... is swollen tissue. (B) is the tip of the ulcer and the opening where the purulent tissue exits.

    An ulcer of five centimeters in diameter has a bag of purulent tissue (C) of one centimeter in diameter. What if the ulcer were larger, say 10 to 20 cm in diameter. In that case, it would be dangerous for us. Skin ulcer has had its development since its inception until the pus.anj formation of the purulent sac (C) and ejection of purulent matter. This process takes place over a period of ten to twelve days. There is no medicine that can stop it or destroy it. There is only medicine that can reduce pain. Let's split the twelve-day period into three phases.

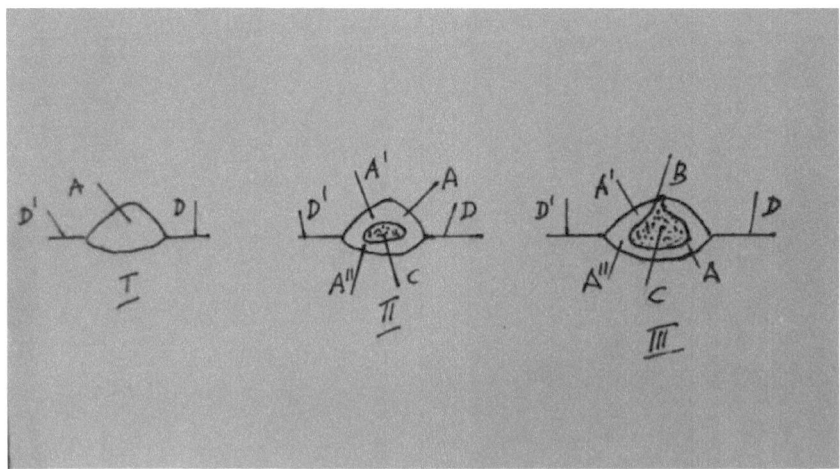

Ulcer development period.

On the sketch, we see three phases of ulcer, which we imagine larger in size than ten inches wide, from its beginning to the end when it ejects purulent secretions. Each phase takes about four days. We imagine that this ulcer is placed in a dangerous position, which can injure the veins, nerves, eye, etc.

What can the doctor do in this case? Can he act in period one (I) and do surgery? He should act in period (I) and remove the middle part of the body of the ulcer, which is defective because no purulent sac has been formed yet.

If the doctor operates in period (II) he could remove the purulent sac (C) and a small amount of tissue around it. In this case, he could save our organs.

In period (III) it is too late for surgery. During this period, our organ passing through the purulent sac (C) was already damaged.

Can we compare this to cancer? There were about twenty-five to thirty students in my class of so called eight-year school. Only a few of us students had a skin ulcer. This means that not all students could get skin ulcer because the composition of our bodies was different.

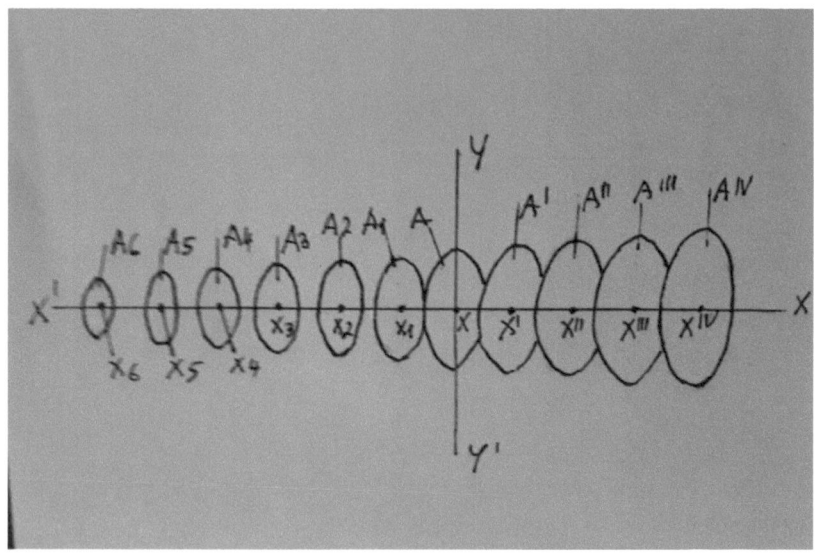

Structure of the disease.

X, X1, X2... Points of the periods from the history of disease.

X', X'', X'''... Points of the periods from the future of illness.

A1, A2, A3 ... Surfaces of history from disease
A', A'', A'''... Surfaces from the future of disease

The sketch shows the body of a cancer-like disease, for example, from its beginning in point (X6). Let's say the cancer started three years ago in position (X6) and the surface (A6) shows a cross section of its body. From point (X) to point (X6) the period of three years is in adequate sections as in points X1, X2, X3 ... On the other side of the diagram are sections of the future of the disease at points X ', X ", X"' ...

The future periods in the diagram need not be equal to the periods in the past. For example, we can only take periods of one month. (A ') is the area on a particular day of the following month. (A ") is the area on a given day after two months or more, etc. Periods can be shorter and longer depending on how we want to study and label the disease.

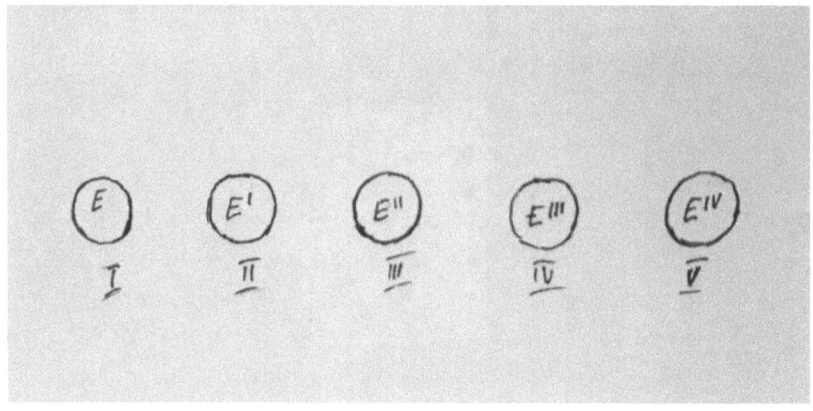

The relationship between Earth precessions and human diseases.

E, E', E"... The position of the Earth on its precession path.

According to my book " Formation Of Planet", Earth does not revolve around the sun. Earth makes certain precession trajectories in the Celestial space. The variability of Earth's position in the Celestial space influences life on it. Earth precessions affect humans, animals, and the entire biological world, and cause disease and their intensity. The Sun is the most important Celestial body affecting the Earth. Solar radiation determines the health of humans, animals and the rest of the biological world on Earth.

Sometimes it happens that fish or animals die for no reason. No one can explain such cases. The reason is actually Earth's position in the Celestial space.

The past sketch shows Earth in five positions in Celestial Space. In Earth's position, one (I) has no cancer disease. In position two (II), cancer is very rare. In position three (III) cancer is common. In position four (IV) cancer is again rare. There is no cancer in position five (V). This is just an idea of how to explain cancer in relation to Earth's positions in Celestial space.

We can ask some questions here. What can we do to prevent ulcers? Was there any medicine for this kind of preventive action? This could have been done in the ulcer season. I actually want to compare this disease to cancer. Some diseases were preventable, such as children drank fish oil to protect them from Rahitis. There is a vaccine we can take as a flu preventative. What can we take to protect us from cancer? Is there medicine or something that can protect us from cancer?

What exactly is the flu? How can we prevent flu? What vaccine will we use?

We often hear that some flu vaccines are not effective. Television often talks about vaccines and their effectiveness. What's really going on?

Flu is actually the result of a celestial effect on the organs of our body. Depending on where the Earth is relative to other celestial bodies, the flu will have such characteristics. This means that each flu is different from one another. In fact, we can define the flu in the following words. Our Earth makes precessions in

Celestial space. Our Earth makes precessions in the Celestial space and changes its position constantly. The influences of Celestial bodies are different every moment. For this reason, every flu is different.

The radiation of Celestial bodies affects the Earth, but since the earth is in a different position at any moment, the impact of radiation will be variable. The power of radiation and the angles of incursion of the Earth will alter the characteristics of the flu. This means that the flu changes every moment of the day, week or month. This way the flu makes the disease mutation all the time.

Taking the flu vaccine can be unsuccessful. In fact, the vaccine is the artificial creation of influenza in a light form to excite the body's resistance. The vaccine should prepare our body for the oncoming flu, in fact, to strengthen our immunity against the disease.

It is not considered here that the flu mutates all the time. In fact, the characteristics of the flu are different every time after the outbreak. The reason the vaccine is not successful is the variability in the characteristics of the flu.

In fact, our body creates the flu based on Celestial influences. In other words, flu is a product of our body.

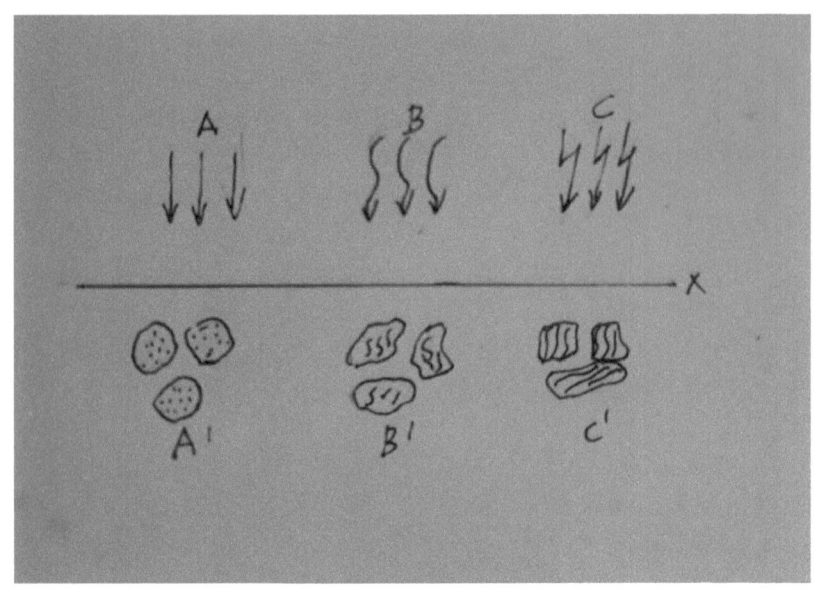

Flu creations.

    A, B, C... External influences.
    X - Line X.
    A', B' C' ... Different body creations created from external influences.

    The sketch shows external influences on our body (A, B, C), which cause the flu to build up in our body (A', B', C'). Each time the external influences are different which will create flu with different characteristics. Flu (A ') has different characteristics than flu (B') or (C '). If we take the flu vaccine (A '), it will not be effective for the flu (B') or (C '). By looking at these creations of our body as the flu, we will understand creations of the skin ulcer, or creations of cancer.
    What are the reasons and effects on our body to create cancer?

Here are some cases when our body will respond to external influences and form diseases:

If one falls into cold water in winter, he will catch a cold. He will cough. His nose will leak. His temperature will rise. The man will feel sick and will lie in bed. Very easily he can get pneumonia and die.

What actually happened in this case?

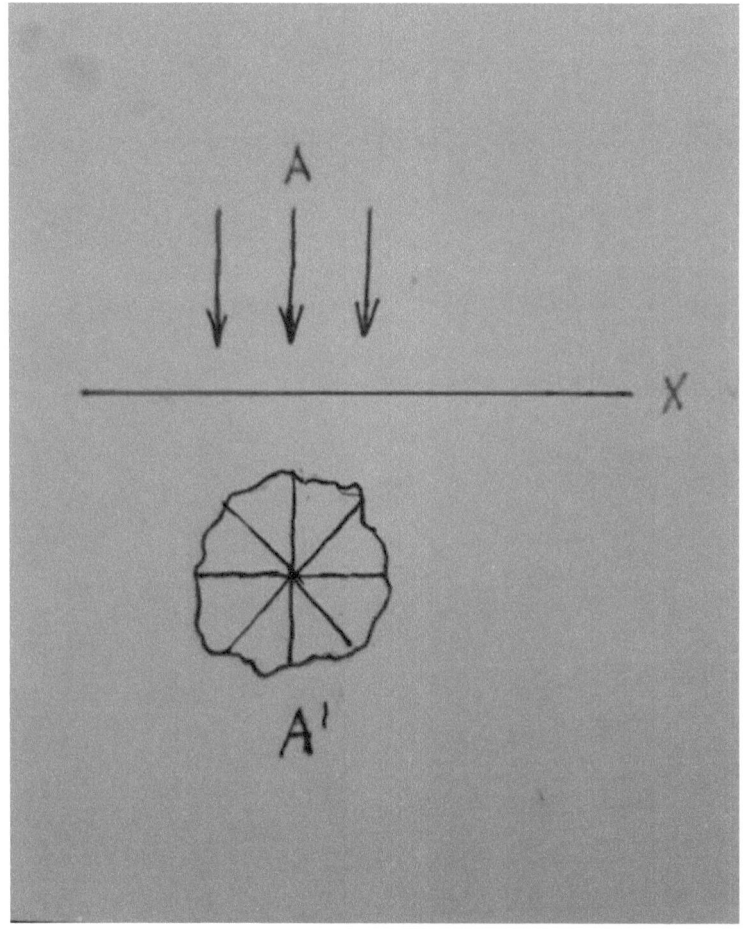

The reason for the disease.

A - External cause of disease.
A' - Disease body section.

    The sketch (A) shows the cause of the disease, falling into the cold water of a river or sea. (A ') is a section of the body of the disease, showing high fever, coughing, sniffling, etc.
    Due to falling into cold water, the human body created elements of the disease, such as high fever, sniffles, coughs, etc. This man can easily get pneumonia and die.
    The human body responds to external influences and creates disease. Many people are sensitive to the impact of some allergy-causing plants.

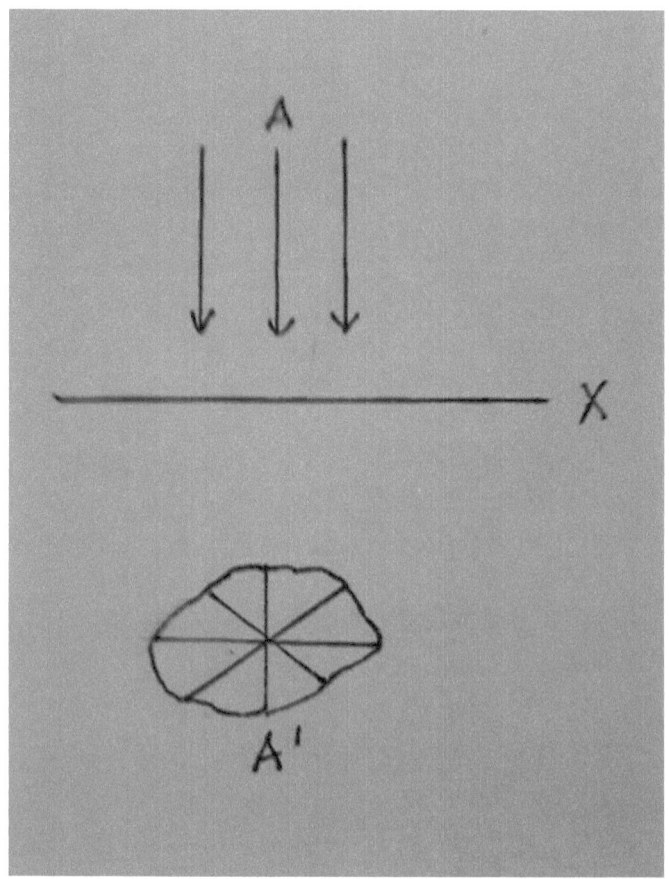

Allergy

A - Impact of plants.
A' - Section from the body of the disease.

In the sketch impact of plants (A) is the cause of the allergy. Some plants release the pollen we breathe. This pollen is the cause of allergy disease. We will cough and sniff. Our body temperature will rise.

In all these examples we have reason (A) and creation of disease (A '). Illness (A ') is actually a creation of our body.

Can we compare cancer to disease in these examples?

Can we accept that the oil, gas, and chemical industries create cancer? In this case external influence (A) is this industry.

This external reason (A), which causes our body to respond and create disease, we can call "Factor X". In the case of cancer, "Factor X" is the oil, gas and chemicals industry. Mostly we believe in this, but we are still looking for clues. We still can't say for sure what is "Factor X" that creates cancer.

Cars release gas that is dangerous to us. Many downtowns have high percentages of gas due to traffic, such as nitrogens, CO2, monoxides, etc.

We may also consider "Factor X" to be of celestial origin.

What is the importance of solar radiation for Earth. The Sun is the most important celestial body for us, whose radiation gives us light and warmth. The variability of the Earth's position and distance from the Sun can cause different effects of radiation.

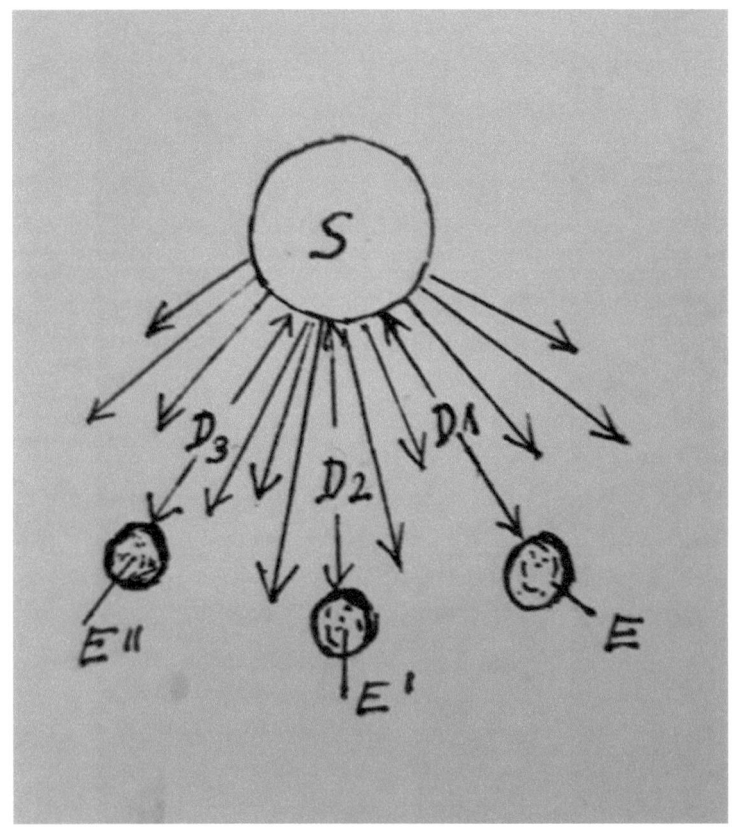

The diversity of Earth's position relative to the Sun.

S – Sun
E, E', E" - Positions from Earth.
D1, D2, D3 - Earth's distances from the sun.

    According to my book, Formation Of Planet, neither Earth nor planets rotate around the Sun, but are stationed in Celestial space. Earth (E, E ', E ") probably makes certain precessions over long periods, and its distances to the Sun (S) are not constant. These displacements are very small that they cannot be drawn to the sketch properly. These displacements are very small that they cannot be drawn on the sketch properly but this is an explanation of the variations of the Sun. The differences in distances (D1, D2, D3) are very small and imperceptible. I just want to demonstrate them here in a sketch.

The sun is the strongest source of radiation for us in celestial space. Earth receives its rays that transform our atmosphere into heat and light. Other planets and stars do not affect us so much, and we can only study the influence of the Sun.

The Earth comes in different positions towards the Sun, as shown in the previous sketch. These variations are small and negligible for solar heat and light, but for the influence of "Factor X" can be considered.

On the other hand, our body lacks some of the elements it needs. We can call these elements "Factor Y". Without these ingredients, our body will not be resistant and our immunity will fall. People have always taken some natural ingredients to strengthen the body against disease. Fish oil was given to children to strengthen their immunity against rickets. In India and the Arab countries, people take a lot of spices like pepper, dried red pepper, etc. Garlic, onions, hot peppers, etc. were taken in our country. All of these food ingredients strengthen our immunity against the oncoming disease.

We will try to replace the loss of "Factor Y" with different spices and medicine. This way we will increase our immunity. Is immunity the only entity that protects our body? I would say no. Increasing immunity will help us a lot, but it will not completely protect us. We can still get sick from an oncoming disease.

****

This is where I described infectious and non-infectious diseases. Infectious diseases live in our body. They use the organs of our body for their own existence. These diseases will reproduce in our body. Such diseases can live on in the human body. They can pass from body to body as infectious diseases.

On the other hand, we have non-communicable diseases. A simple example is an injury to our arm.

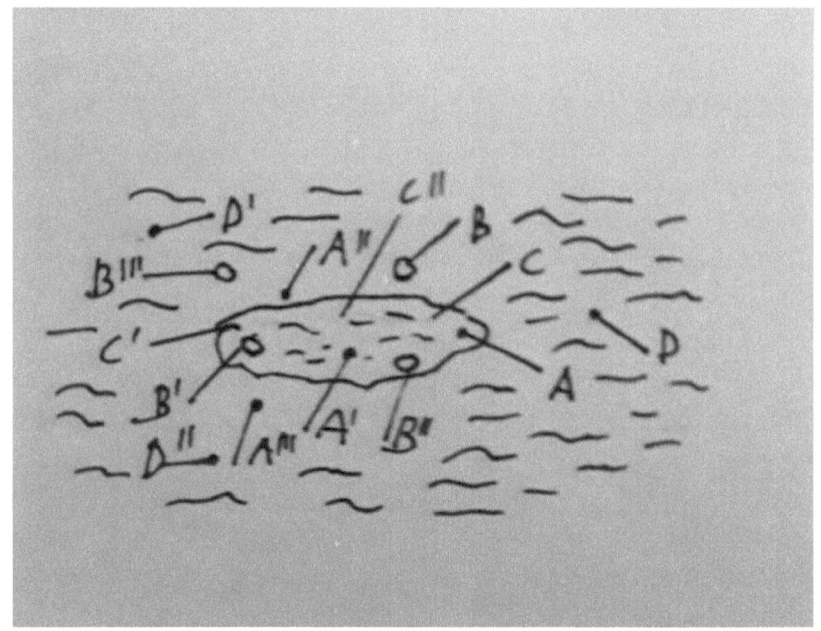

Microorganisms in the wound area.

A, A', A" ….. Viruses in the wound area.

B, B', B" ….. Bacteria in the wound area.
C, C', C"…. The area of the wound.
D, D', D"…. Skin

This sketch shows the wound area (C). D, D ', D "... is the skin around the wound area. In this area we will be able to find many microorganisms as bacteria and viruses. Why are these microorganisms in the wound area? How did they get there? Today's science will say that they came from the air mass around us. Similar explanations are incorrect. These viruses and bacteria are created by our body. They are a product of our tissue.

When a person gets sick, he dies. Way? His organs are no longer functioning as they were before. His nerves weakened. His blood vessels are not as good as they used to be and they cannot circulate blood as before. The heart will not be able to endure. The man will die.

In this case, the man died even though he was not ill. The reason for his death is age. His old body was no longer functioning. His heart stopped. We can say that the man died of a heart attack. One always dies for heart failure, whether ill or healthy.

What happens to cancer? How long will cancer last? I mentioned before that the ulcer of larger dimensions disappeared. Today, there are only small skin acnes. Will the ulcer reappear? Will its diameter reach six or more centimeters. I wonder if ulcer still exists in our body in some of its tiny form.

We can definitely say that ulcer can change its configuration. It can get small or big depending on the celestial season, but it will act like an ulcer. It can completely disappear and reappear. Cancer can also take many forms depending on the celestial season. It may disappear or take on another form. Will cancer live on forever in the human body by changing its shape from small and harmless to bigger and dangerous.

We came up with a cancer mutation. I wonder how many mutations cancer has had throughout history. This means that cancer has been a different mutation of the disease in the past.

This means that we cannot know what our body is hiding. What kinds of diseases are diminished but still exist in us. Will they expand in the future? We can put this opinion in another way. Depending on the celestial season, our body will create a certain disease.

In the celestial ulcer season, our body has been creating this disease. The celestial season depends on relationship between Earth and Sun, in addition on the relationship between Earth and Moon, and other celestial bodies that can influence Earth through its light emission.

Now that the earth is in a definite celestial season, which we can call "Cancer Season", the body will produce cancer.

Many diseases have disappeared as their seasons have passed. As a kid, I had mumps which was quite a painful disease.

What is the reason that the body creates the aforementioned diseases. These diseases are the cause of certain celestial influences.

We can once again consider a simple example, when a person falls into cold water. The mucous membranes of his mouth, nose and throat will inflame. The man will cough. His nose will drool. The temperature of his body will rise. There will be a risk of pneumonia.

What actually happened to the body?

Falling into cold water is actually a shock. At first the nerves will register a shock. They will respond and transmit the shock notification to the organs of the body. Soon the body will react and the temperature will rise.

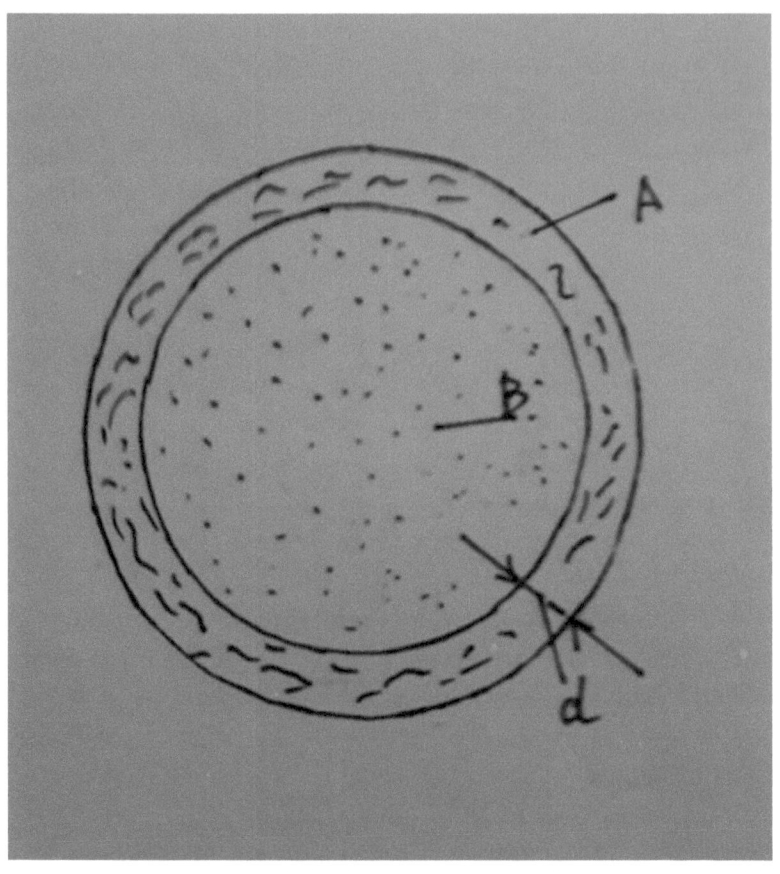

Injury to the body due to falling into cold water (1).

A - Body injury area.
B - Remaining healthy body area.
d - Percentage of disease .

    On this sketch we can see a deseased body size (A), if one falls into cold water. Area (B) is the remaining healthy area.
    If a man falls in cold water of ten to twelve degrees his body will fall ill. Depending on how long he has been in the

water, the disease will be weaker, or worse, as well a person may die.

In the upper sketch, the human body is diseased from 15 to 20 percent. This is shown by percentage of disease (d). This sketch can be drawn by the doctor every four to six hours, so it can be estimated at what risk the patient is. Of course this will be done based on medical parameters, such as temperature, pulse, sweating, vomiting, etc.

This way we can demonstrate human disease. According to the diagram we can determine wheather a person will die or survive.

Injury to the body due to falling into cold water (2).

A - Body injury area.
B - Remaining healthy body area.
d - Percentage of disease.

In sketch (2) area (A) is larger than area (B). The percentage (d) is greater than 60 percent, and most likely one will not survive.

Due to ilness in falling in falling in cold water, the body will form different disease parameters. First the body temperature will rise. The following mucous membranes of the mouth, nose and other airways will swell and inflame. Way...?

All body functions are nerve controlled. In this case, nerve control was reduced or completely scheduled.

The man will fight for life under the influence of disease. He will cough. His nose will leak. All these diseases are transmited by the body. The patient will get pneumonia as the next disease and he will die. His heart will stop.

Falling into cold water caused injuries to the body and caused several more diseases. All of this caused the death of the patient. Can we compare this case with cancer?

The fall into cold water we can indentify. We can describe all stages of the disease that will result from this. We can scientifically investigate all parts of this case and its consequences.

Skin ulcer was conceived somewhere in our body. We need to determine where it belongs. Can we assume the onset of skin ulcer in our body?

The first organs to feel foreign influence are our nerves, or skin, for example. Our nerves will transmit signals to our brain, which will try to understand the impact. Our brain can easily detect cold, heat, humidity, etc. I am thinking, how our brain can recognize celestial influences. Any influence will in some way trigger the reaction of our body. Suppose that cold weather will cause colds, sores, runny nose, fever, etc. Celestial effects will cause skin ulcer, cancer, etc.

In the case of cold, nerves send information to the brain, that our skin feels cold of the weather, and we will feel the cold as the definition of our brain. Our mucous membranes will turn red. It will swel and body temperature will rise. In this case, the brain follows the activities that take place in our body. Our nerves are trying to involve all organs in the fightagainst cold. The body will come into dysorder. Our heart will push the bloode harder. We are actually fighting an enemy called cold. If the weather is too cold, the human organs will not win and the human will die.

In this case the enemy was discovered, ie cold weather. In the case of cancer, we cannot say for sure way it was created. In other words, we cannot say way cancer originated in one's body. Cancer is celestial disease whose cause we hardly recognize with current knowledge.

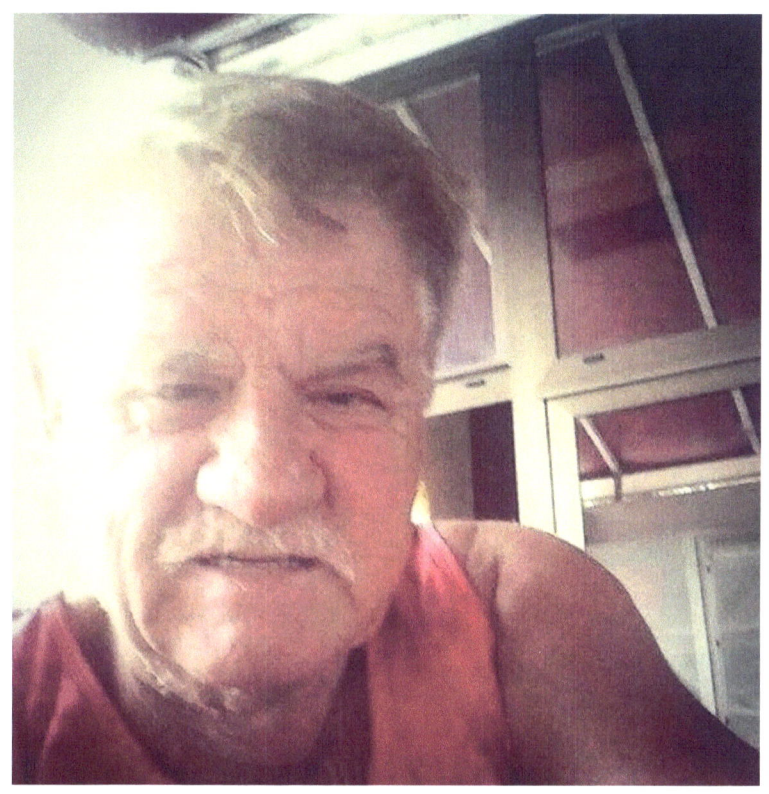

PERO METKOVIC

I was born in Molunat, a fishing village in Croatia.
First I wrote stories and poetry in Croatian language.
I later published the book "Formation of Planet".
It describes my vision of astronomy that is different
from existing science.
I published a thriller "Tattoo of Dragon", which describes a girl
fighting a gang of criminals for the police.
I have released two screenplays, " Spider X " and " War Story ".
I was a sailor, officer, and later a captain at ships of the merchant

navy. I have been traveling around the world many times during my life.

I hope this book will be my contribution to the world science.

www.ingramcontent.com/pod-product-compliance
Lightning Source LLC
Chambersburg PA
CBHW040221220526
45473CB00001B/72